IVF: A Patient's Guide

REBECCA MATTHEWS, PhD

Published by Get It Right, Inc. Portland, Oregon

Book Design by Joshua Matthews

Photographs and images used with permission, courtesy of Oregon Reproductive Medicine, Portland, Oregon, and Reprogenetics, LLC, New Jersey

ISBN 978-0-557-73100-8

Dedicated to the patients of Oregon Reproductive Medicine.

Disclaimer

The contents and information in this book are for your informational use only. This book provides general health and fertility information and is not intended to be a substitute for professional medical advice, diagnosis or treatment. Always seek the advice of your physician or other qualified health provider with any questions you may have regarding a medical condition. Never disregard professional medical advice or delay in seeking it because of something you read in this book.

Acknowledgements

Without the support and guidance of Alison Coates, I may never have become an embryologist much less produced this work. Thank you to everyone who had a hand in the editing process including Courtney Sheehan and my editor in chief and husband Josh. Collin McKean, Esq. for his legal counsel and contribution. The patients, so willing to share their experience and insight to aid others who find themselves in similar circumstances deserve a special level of gratitude. Finally, a big thank you goes to, Drs. Hesla, Matteri, Bankowski and Barbieri and the entire staff and management of Oregon Reproductive Medicine for their support, their encouragement and for their constant devotion to deliver miracles to the arms of hopeful parents all these years.

Preface

Couples pursuing fertility treatments have often tried to conceive for months or even years without success. This can leave them with the burden of disappointment and with feelings of depression, anger and resentment. The decision to visit a fertility specialist can initiate contrasting feelings of hope and fear and can present tremendous emotional and financial stress. It is important to recognize these feelings and despite the grief you may feel, realize that there are reasons to be positive about the journey ahead.

Here are some ways to help you prepare for fertility treatments:

1. Gather information in advance
2. Prepare for decision making
3. Look after your emotional well being and your relationships
4. Set up a strong support system
5. Identify sources of stress and coping mechanisms
6. Decide what you have control over and what you don't
7. Anticipate the unexpected
8. Look ahead and minimize regrets

Loss of control and fear of the unknown are common feelings. This book was written to empower you to know what questions to ask, what decisions need to be made and how you can control your own destiny. It is my hope that this knowledge reduces the stress of fertility treatment and hopefully gets you closer to your dream of becoming pregnant.

I wish you all success in your journey.

Table of Contents

1

Choosing a Fertility Center

You've been trying to get pregnant for a while without success; what do you do now? Couples who have tried for more than a year and women over age 35 who have tried for more than six months should seek medical advice to help them on their path to parenthood. Most people begin the process with a visit to their obstetrician/gynecologist (OB/GYN) who is trained to diagnose and treat general reproductive disorders and may also be able to provide limited fertility treatments. Although your OB/GYN is not trained in more advanced fertility treatments, they can refer you for testing and advise the best course of action based on your medical history.

Your doctor might suggest you see a Reproductive Endocrinologist, also known as an RE. These are doctors who specialize in treating infertility and can treat both men and women. In the USA, there are approximately 800 board certified REs and 400 IVF clinics. How do you know which one is right for you?

Choosing a fertility clinic is one of the most important steps in your treatment process and has a major impact on your chance of success. It is very important to choose carefully.

Considerations

- Location and accessibility
- Number of doctors
- Credentials, experience and personality of staff
- Services offered
- Cost, insurance and payment options
- Reputation and references

- Patient education
- Waiting list
- Support and amiability of staff
- Statistics for your age group
- Overall multiple pregnancy rates
- Donor egg statistics

LOCATION

The stress and inconvenience of fertility treatments can be eased if you choose to go to a clinic close to home. You may require an IVF cycle, during which there will be daily visits for blood tests and ultrasounds. A two-hour drive each way to the clinic would definitely make this a more stressful experience. On the other hand, you don't want to compromise your chances by choosing the most convenient location if their statistics are below average.

The key here is to find the best clinic nearby. The Society for Assisted Reproductive Technology (SART) has created a website where you can search for a clinic within a certain mile radius of your zip code: www.sart.com. This is a good place to start. Decide how far you would be willing to travel. Could you get time off work if needed and do your finances allow hotel stays and even airfares?

Some people travel far and wide to get the best care, especially if they need special treatments such as a gestational carrier or donor egg IVF. If you are starting this journey and need a diagnosis and possibly low tech treatments then it may be a good idea to choose a doctor in your town. Ask your OB/GYN or family doctor to recommend two or three fertility doctors and then research those clinicians before you make a final decision. This research should include discovering what services they offer, reviewing the pregnancy rates and obtaining patient references.

ACCREDITATIONS OF DOCTORS AND CLINICS

Reproductive Endocrinologists are devoted specifically to treating infertility. They have completed a two to three year fellowship in infertility treatments, followed by two years of clinical experience. Board certified RE's have also taken written and oral exams and are certified by the

American Board of Medical Specialties. Ask your doctor if he or she is board certified or board eligible.

Remember, having these credentials is just the first step; many factors make a good doctor. Most important is whether you are comfortable asking questions, sharing information and feel like a partner in the decision making aspects of your treatment.

The Joint Commission, CAP, or a similar organization should accredit the embryology lab. These organizations ensure the highest quality of healthcare standards are met and maintained. The embryology staff should be well practiced in their techniques and the lab should be sufficiently staffed to handle the workload.

SPECIALTIES AND SERVICES OFFERED

Each clinic will offer a range of services, which usually includes infertility-testing, diagnosis, IUI and IVF. Some of the bigger clinics have very specialized treatments which are individualized for each patient while smaller clinics may have one or two simple treatment protocols that are used for everyone. There are pros and cons to both large and small clinics. Depending on your situation you may require a simple or more advanced intervention. Ask yourself:

- Do they offer a full range of services?
- Do you feel that the doctor is performing adequate testing to diagnose the problems before suggesting a solution?
- Do they allow experimental or alternative medicine to be used in conjunction with your traditional treatment?

Beware of programs offering new, untested technology or those with promises that seem too good to be true. On the other hand you want to make sure that the clinic has kept abreast of changes in the field and offers the most up-to-date, tried and tested technology available.

If you are doing IVF, an important component to look for is whether the embryology lab offers extended embryo culture to the blastocyst stage. This is where embryos are grown in the lab for five days before transfer. Extended embryo culture has several advantages; it allows better selection of embryos, reduces the number transferred and can increase pregnancy rates. Although clinics have their own criteria for who qualifies for blastocyst transfer, be wary of a clinic that discourages transferring the embryos at the blastocyst stage; the lab may not have the capability of supporting good embryo growth.

Another important advance in recent years has been the use of a new freezing method called vitrification for the cryopreservation of eggs and embryos. Many studies have shown advantages of vitrification over the older methods of freezing. Although this shouldn't be the deciding factor; whether or not the embryology lab does vitrification can indicate how current they are with their techniques.

Some clinics have special expertise. This might include treating women with diminished ovarian reserve, gestational carriers, genetic testing of embryos, blastocyst culture and donor egg IVF. Most important, you need to know a clinic is able to offer all the services you may require.

SIZE OF CLINIC: NUMBER OF CYCLES ANNUALLY AND NUMBER OF DOCTORS

A solo practitioner may operate a clinic while others have five or six doctors or more. Regardless of the size, be sure you are getting the quality of care you desire and deserve.

Accessibility to staff, availability of doctors out of hours and speed of response to queries are all important factors that can minimize frustrations during fertility treatments. Find out who is available to answer your questions and concerns when the clinic is closed and what happens if your treatment falls on the weekend.

Consider the doctor's experience and how long they have been practicing various treatments and techniques. If a clinic has several REs, you might find that they are on a rotation and you won't necessarily see your own doctor for every appointment. Some people prefer to see the same doctor every time in which case a smaller clinic might be better for you.

The initial friendliness and helpfulness of the staff and the time taken to answer questions will give you a feel for what kind of care you can expect from a clinic.

Look at the number of cycles performed each year (you can find this on the SART website). Some "mega-clinics" perform thousands of cycles a year and others just a handful. The bigger clinics will most likely run with the efficiency of a well-oiled machine but you may end up feeling like a number with a revolving door of doctors. Statistics show that the size of the clinic (measured by the number of patients treated per year) does not affect the average success rates.

Everyone is different and each person needs a different level of support from his or her clinician. Think about what kind of clinic you would like to go to and then find the best one that offers what you are looking for.

REFERRALS FROM DOCTORS AND OTHER PATIENTS

The reputation of the clinic is one of the best indicators of what you can expect. If an OB/GYN refers someone to a fertility clinic, they will most likely see that person again if they become pregnant. Ask your OB/GYN if most of their referrals return for obstetrical care and what the patients had to say about their experience at the fertility clinic.

You can ask an IVF clinic if any previous patients are willing to share their experiences with you. An alternative is to go to Internet bulletin boards and community websites. A good one is www.ivfconnections.com.

WARNING

Do not trust the medical advice given on these boards. Only trust your healthcare professional.

I have heard a lot of unsolicited advice given and some of it can be dangerous. In my opinion, the Internet is a good place for emotional support and to share experiences. If you are choosing a clinic, you should hear what the *majority* of people are saying. There will always be someone who has a bad experience. Don't let one or two negative comments put you off if everyone else is singing their praises.

STATISTICS

Clinics are required by law to report their annual in vitro fertilization (IVF) cycles and subsequent pregnancies to the Center for Disease Control (CDC). These figures are published each year by an organization called the Society for Assisted Reproductive Technology (SART). The figures are published two years after the cycles were initiated because live births have to be recorded. Therefore, for example, the figures for 2008 are published in 2010.

A full list of pregnancy rates from centers across the country can be viewed at the CDC website or at www.sart.com.

Here is an example of the data available for each IVF clinic; these are the average success rates from 2008 according to SART.

PREGNANCY SUCCESS RATES

Type of Cycle	**Age of Woman**				
	<35	**35–37**	**38–40**	**41–42**	**43–44**[d]
Fresh Embryos from Non-donor Eggs					
Number of cycles	39621	21744	20430	9243	6152
Percentage of cycles resulting in pregnancies[b]	47.6	38.0	30.3	20.4	8.8
Percentage of cycles resulting in live births[b,c]	41.3	31.1	22.2	12.3	4.1
Reliability range	40.8–41.7	30.5–31.7	21.6–22.8	11.6–12.9	3.6–4.6
Percentage of retrievals resulting in live births[b,c]	44.5	34.9	26.1	14.9	5.2
Percentage of transfers resulting in live births[b,c]	47.3	37.3	28.2	16.7	6.8
Percentage of cycles with elective single embryo transfer[b]	5.2	3.2	1.0	0.5	0.3
Percentage of cancellations[b]	7.2	10.9	14.8	17.8	21.1
Average number of embryos transferred	34.1	24.8	16.7	9.3	4.0
Percentage of pregnancies with twins[b]	2.2	2.4	2.7	3.1	3.3
Percentage of pregnancies with triplets or more[b]	33.3	28.1	23.5	15.4	11.4
Percentage of live births having multiple infants[b,c]	1.9	2.0	1.7	0.6	0.4
Frozen Embryos from Non-donor Eggs					
Number of transfers	10303	5382	3639	1194	920
Percentage of transfers resulting in live births[b,c]	35.6	29.5	26.1	19.3	13.2
Average number of embryos transferred	2.2	2.1	2.3	2.3	2.4

PREGNANCY SUCCESS RATES, continued

All Ages Combined[e]

Donor Eggs	**Fresh Embryos**	**Frozen Embryos**
Number of transfers	9905	5319
Percentage of transfers resulting in live births[b,c]	55.0	32.7
Average number of embryos transferred	2.1	2.2

[a] Reflects patient and treatment characteristics of ART cycles performed in 2007 using fresh non donor eggs or embryos. [b] When fewer than 20 cycles are reported in an age category, rates are shown as a fraction and confidence intervals are not given. Calculating percentages from fractions may be misleading and is not encouraged. [c] A multiple-infant birth is counted as one live birth. [d] Clinic-specific outcome rates for women older than 44 undergoing ART cycles using fresh or frozen embryos with non-donor eggs are not included because of small numbers. Readers are urged to review national outcomes for these age groups. [e] All ages (including ages > 44) are reported together because previous data show that patient age does not materially affect success with donor eggs.

While helpful, the reports can be a little confusing. Keep in mind that a "cycle" begins when the female patient starts taking the drugs that stimulate the production of multiple eggs in the ovary. The IVF clinic reports each cycle to the CDC within a few days of the medication being started, this ensures that all patients are reported and not just the ones who respond well.

Around 10 percent of all the patients starting to take stimulation drugs will be cancelled before the day of egg collection, usually due to a poor response. Approximately 10 percent of patients reaching egg collection will not have an embryo transfer for the following reasons:

- Around 1% of all egg collections will have a complete failure of fertilization and therefore there will be no embryos to replace.
- Some patients produce too many eggs and can be at risk for ovarian hyper-stimulation syndrome. We would not transfer any embryos during this cycle because pregnancy would exacerbate the condition. We would freeze all the fertilized eggs and replace them in a frozen embryo replacement cycle at a later date.
- Some patients do not develop a suitable lining in their uterus during stimulation and therefore we would freeze any fertilized eggs for use in a future cycle.

Pregnancy rates are quoted from three stages of treatment, cycle start, egg retrieval and embryo transfer. The lowest rate will be from the start of stimulation and the highest rate will be from the embryo transfer stage. Because a pregnancy can only be achieved when embryos are placed in the uterus, the pregnancy rate from embryo transfer is a good place statistic to compare clinic to clinic.

Statistics for Donor Age Group

Look at the statistics for women in your age group and compare this to the national average (the most recent average success rates are available on the SART website and the specific clinic website might have more recently updated statistics). It is wise to use a clinic that has at least the national average success for your age. Remember that many factors affect success rates, some clinics may refuse to treat women with a low chance of success, whereas others may accept all patients. This could inflate the success rates of some clinics and lower the rates of others when only a small number of people are treated each year. The SART data is, however, a good starting point to compare the success rates of clinics.

Use the SART success rates as a piece of the puzzle; your clinician should be able to give you a personalized chance of success following diagnostic testing and a review of your medical history.

Statistics for Donor Egg Group

The great equalizer of all clinics' statistics is the results of donor egg IVF cycles. These cycles involve the donation of eggs from young women who are specially selected based on their normal fertility.[1] Most clinics have strict rules as to whom they allow to be egg donors and so these success rates should be very high.

The most predominant factor when predicting the outcome of an IVF cycle is the age of the egg, specifically, the age of the woman from whom the egg came. If the eggs are from women in their early twenties, who have

---

[1] The donor egg rates reported to SART also include the rates for known donors who are friends, sisters and relatives of women choosing IVF. These egg donors are selected for personal reasons and not because of their fertility and so have the potential to slightly lower the egg donor rate for clinics that carry out a lot of known donor cycles. If this is the case, ask the clinic for their statistic for anonymous egg donor cycles only.

been through an intensive screening process, the success rates should be close to 100%. We know that many other factors affect the outcome such as the male genetic contribution and uterine factors, and so the success rate is never 100% but in top clinics, the live birth rates from donor egg IVF is close to 80%.

The donor egg live birth rate is a measure of how effective the ovarian stimulation protocols are and how well luteal support regimes work. More important, it is a measure of the lab and their ability to culture embryos and select the "right" ones for transfer. *A low donor egg rate should be a cause for concern.*

MULTIPLE PREGNANCY RATES

On the SART report for each clinic you will notice the average number of embryos transferred, the number of twin pregnancies and the number of triplets for each age group. Look at the under 35 age group. The average number of embryos transferred should be close to 2, the triplet rate should be low (less than 10% and closer to 5%). If this is the case and the pregnancy rates are above average, it is a sign of a good clinic.

Look at the rates for your age group. What is the average number of embryos transferred and what are the multiple pregnancy rates? Does this seem like a reasonable risk to you? How does it compare with other clinics you are considering?

COST AND FUNDING OPTIONS

Most fertility treatments are not covered by insurance, although some of the testing and diagnosis may be. Find out if a clinic takes your insurance for any part of the treatments and what funding options they have. A number of clinics work with financers who offer special loans for fertility treatments with affordable monthly payments.

Some clinics offer plans whereby the patient pays for a "guarantee" of success. For example, a flat fee of $20,000 is paid before treatment starts. This entitles you to a maximum of three cycles of IVF and frozen embryo transfers until you become pregnant. Women who become pregnant on the first cycle will pay more than the usual fee-for-service and people who need two or three cycles will pay less per cycle. These programs "share" the cost of treatments between the participants. If a woman fails to get pregnant through all the treatment cycles, a full refund is given.

To qualify for this type of program, the patients have to fulfill various criteria, which would give them a good chance of success from the outset.

RESTRICTIONS

Some clinics have restrictions about whom they will treat. The restrictions can include age, body mass index (BMI), FSH levels (measuring ovarian reserve) and other medical conditions. If you have any unusual conditions, which you feel may exclude you from being treated, talk to your doctor about the risks involved and why this decision is being made. Is there anything you can do to change the condition or are there any clinics that will treat you regardless? Contacting a number of clinics would be recommended.

The American Society for Reproductive Medicine (ASRM) guidelines should be followed by a clinic to ensure the safest possible outcome and the maximum chance of success.

ALTERNATIVE THERAPIES

Acupuncture has been shown to be beneficial for women undergoing fertility treatments and may increase the chances of success. Does your clinic allow an acupuncturist to treat you during your embryo transfer and do they support alternative therapies, which may help to provide stress relief during the process? Some fertility clinics are affiliated with wellness centers that provide a range of alternative therapies and a holistic approach to your well-being.

PATIENT SUPPORT GROUPS AND MENTAL HEALTH SUPPORT

Fertility treatment is an emotional process. It helps to advocate for your own care and become empowered with information and support. Most clinics offer psychological counseling for any patient who requests it, or they can recommend a good outside psychologist. Many clinics offer support groups enabling patients undergoing the process to meet and talk in a safe and confidential environment with others facing similar challenges.

A good place for information is the group RESOLVE; an organization devoted solely to the needs of infertile men and women. They offer resources to help people understand their options and become connected with others who are experiencing infertility. The American Society for Reproduction (ASRM) is also a great resource for information on various procedures as well as the latest guidelines for fertility treatments.

Regularly discuss with your partner whether you both feel comfortable with the care you are receiving and that you understand all your options. If you feel your needs are not being met, first talk with your RE or clinic director. Don't forget that you can change doctors or clinics anytime.

Ultimately, you are the most important person in this process and you must do what is right for you. This may involve changing doctors within the same practice or moving to a new clinic altogether. Don't be afraid of hurt feelings; this is a business and most RE's will understand the reasons behind a switch. People come, people go (and some people come back again!) always with the same goal in mind; finding the best match for them and of course…getting pregnant.

2

Diagnosis and Treatment

When you first visit your chosen clinic, they will begin by taking a detailed medical history to determine the probable cause of your infertility. Diagnostic tests follow and can include blood work, ultrasounds and possibly more invasive tests that look inside the body. In this chapter we will look at the various tests the doctor may perform, diagnoses and the treatment options available.

In order for you to become pregnant naturally, a series of events must occur in a specific order. If any one of these fails, you will not be able to conceive.

To become pregnant:

- You must have a reserve of good quality eggs
- The egg must be ovulated each month
- The fallopian tubes have to be open and clear of blockage
- The sperm count has to be high and the quality normal
- The egg and sperm must bind together
- The embryo has to develop and be genetically normal
- The uterus must be able to accept the developing embryo
- Hormone levels have to by sufficiently high to support the pregnancy

The doctor will test for as many of these components as possible before deciding what treatment is the best option.

The following graph shows the general distribution of infertility causes. This data was reported to the CDC for all the IVF cycles carried out in an average year.

CAUSES OF INFERTILITY

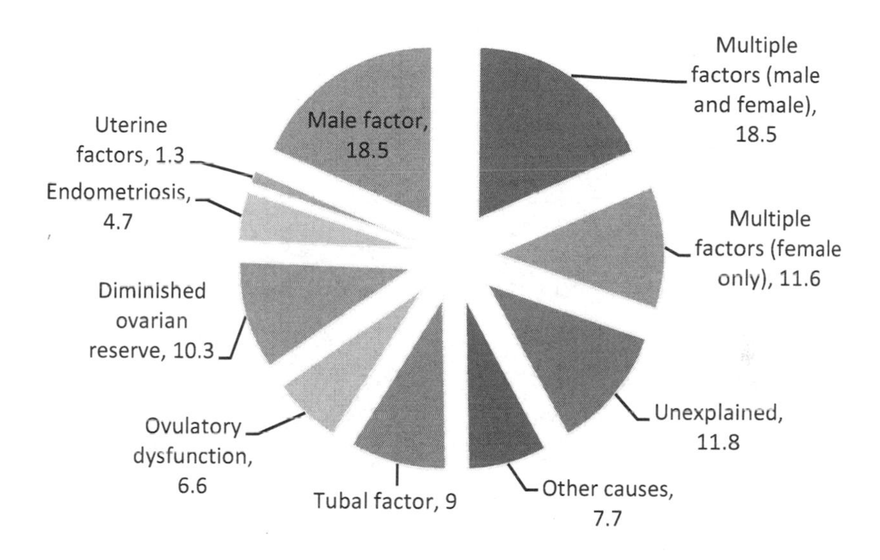

Tests for Diagnosis of Female Infertility

Diagnosis almost always begins with a detailed medical history. By asking questions about your medical, surgical, gynecological and obstetric history, doctors can discover relevant information, which may explain or provide a clue as to why you aren't getting pregnant. This is followed by a pelvic ultrasound examination that can reveal abnormalities of the uterus, fallopian tubes or ovaries. The number of resting follicles in the ovary can also be counted using ultrasound to give an indication of the potential response to ovarian stimulation during a treatment cycle.

Several blood tests are usually carried out early in your diagnosis phase.

These may include some of the following:

- Follicle Stimulating Hormone (FSH)
- Luteinizing Hormone (LH)
- Prolactin
- Estrogen
- Progesterone
- Testosterone
- Thyroxin
- Thyroid stimulating hormone

The body is a delicate balance of chemicals and these blood tests help identify whether there is an imbalance in the endocrine (hormone) system that could be contributing to your infertility.

BLOOD TESTS TO MEASURE OVARIAN RESERVE

Several blood tests can be used to assess how many eggs are left in the ovary (also called ovarian reserve). The tests include Anti-Mullerian Hormone (AMH), Day 3 FSH and a Clomiphene Challenge Test (CCT). These are strong predictors of how well your body will respond to FSH stimulation. The results of these blood tests along with the resting follicle count (also called the antral follicle count) are the most conclusive tests of ovarian reserve.

A measure of the FSH level shows how hard your body has to "work" to stimulate a follicle to grow each month. A high FSH level on day 3 shows that the pituitary gland in the brain is trying to stimulate an ovary that has a diminished capacity to respond. This is a poor prognostic sign.

Although FSH level can correlate to both the quantity and quality of eggs, it should be taken as part of the bigger picture and used in conjunction with other diagnostic factors. The age of the woman is the most overwhelming predictor of success from fertility treatments. A young women with a high FSH still has a better chance of becoming pregnant than an older woman with normal FSH. The older woman may have many eggs but they are more likely to be abnormal because of her age. In the end, quality is what really counts.

FSH levels can predict the chance of the cycle being canceled due to a low response or low egg yield, while age is a better predictor for pregnancy. Ovarian reserves, together with the female partner's age are the best predictors of a treatments success.

CLOMID CHALLENGE TEST (CCT)

For the CCT, blood is drawn on day three, four or five of the menstrual cycle and the levels of estrogen and FSH measured. The woman will then take Clomid tablets (also known as Clomiphene Citrate) between days five and nine of the cycle. Blood is drawn around day 10 of the cycle and the FSH level measured once again. The highest level of FSH, whether it is the day three or the day 10 level, is used as the test result.

The FSH level following Clomid indicates how well a woman will respond to fertility drugs and her likelihood of conception, depending on age. In general an FSH level over 10 can indicate a lower than average success rate. A level between 12 and 14 the chances of pregnancy are further reduced and with an FSH over 15 the chances may be lower still.

There are many cases of women with elevated FSH levels becoming pregnant and having healthy babies. If you are found to have an elevated level, don't be too disheartened. Work with your RE to find the best treatment suited to you. Most doctors will try at least one IVF cycle with your own eggs. You may just need a higher than average dose of stimulation drugs depending on your response to Clomid. Another option is to do a "mini-IVF" cycle with Clomid or Letrozole. Experience has shown that women who are only expected to get two or three eggs with high doses of injectable gonadotrophins (stimulation medication) can do just as well with Clomid or Letrozole at a much reduced cost.

ANTI-MULLARIAN HORMONE (AMH) TEST

AMH is a protein, which is made by cells in small ovarian follicles. Production is highest in the early stages of follicular development, when the follicles are less than 4mm in diameter, and stops when the follicle gets bigger. There is almost no AMH made in follicles over 8 mm in size. Because only small follicles only produce AMH, the circulating blood level can be used as a fairly accurate indicator of how many tiny microscopic follicles are left in the ovary. This is the ovarian reserve.

With increasing age, the ovarian reserve declines and so does the amount of AMH produced. On the other hand, women with many small follicles (such as those with polycystic ovaries) or a good supply of primary follicles, have a high level of AMH.

Remember, AMH level is not an indicator of egg *quality*. We do know, however, that the more eggs we collect, the greater chance we have to create good quality embryos for transfer. Therefore, AMH levels may give us an indirect prediction of a woman's chance to conceive.

One advantage AMH testing has over FSH testing is that the levels are quite constant and testing can be done on any day of the cycle. No test is perfect and so the results from both the FSH and/or AMH testing are used in conjunction with other factors when predicting success rates for fertility treatments.

SUMMARY OF OVARIAN RESERVE TESTING

- Egg quantity *and* quality decline significantly as women age. Both quality and quantity can be average for her age, better than average, or poorer than average in an individual
- Tests for FSH/AMH indicate ovarian reserve and not the quality of the eggs
- Younger women are more likely to have a higher percentage of normal eggs, even with low ovarian reserve
- Older women are more likely to have a higher percentage of abnormal eggs, even with normal ovarian reserve
- The egg naturally recruited by the ovary each month is random; good quality eggs are not more likely to grow than poor quality
- The eggs retrieved for IVF are random; good quality eggs are not more likely to be retrieved than poor quality

Most couples will have basic fertility testing done before being informed of their chance of success and treatment recommendations. Basic fertility testing usually includes:

- Measurement of ovarian reserves (eg. FSH level, CCT, Resting follicle count, AMA)
- Hystersalpingogram to check the uterus and fallopian tubes
- Semen analysis

It is possible that following these tests, no further testing will be necessary. Pregnancy can be attempted during the diagnostic phase; for example, an IUI can be done in conjunction with a Clomiphene challenge test and can be carried out in the same cycle as a hysterosalpingogram. These basic tests can be started as soon as you visit the fertility clinic.

Here are some examples of further tests, which may be performed depending on your situation.

HYSTERSALPINGOGRAM (HSG)

A hysterosalpingogram (HSG) is a common test used to determine whether the tubes are open (patent) and if the uterine cavity is normal. During an HSG, a catheter is placed through the cervix into the uterus and a contrasting dye is injected into the uterine cavity. Several x-rays are taken of the pelvic area in order to identify if the dye is travelling through the tubes, indicating that they are clear, and whether there are any uterine abnormalities. An HSG is preferable to a sonohistogram (see below) as it provides more detailed information about the reproductive system, including the fallopian tubes. An HSG is part of the basic testing done in the early stages of diagnosis, it is performed between days 5 and 11 of the menstrual cycle and a woman can attempt conception in the same month.

An x-ray taken during an HSG:

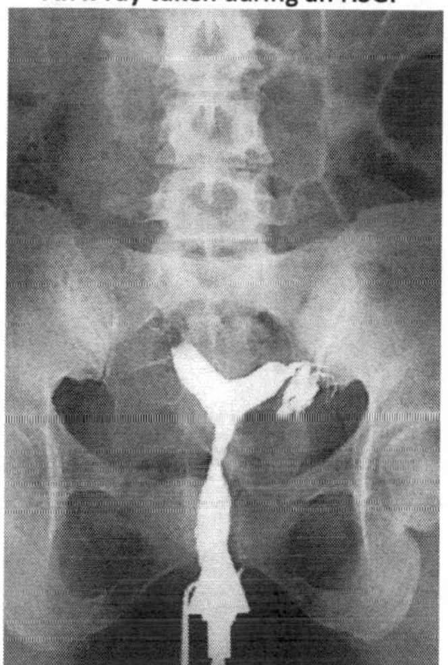

This test is typically performed by a radiologist in the X-ray department of a hospital or clinic and usually takes 15 to 30 minutes. You may feel some cramping similar to menstrual cramps during the test and for a short time after.

SONOHISTOGRAM, HYSTEROSONOGRAM OR SALINE INFUSED SONOHISTOGRAM (SIS)

A sonohistogram is similar to an HSG, except it uses ultrasound instead of x-ray to visualize the uterine cavity. A catheter is used to insert saline into the uterine cavity and then an ultrasound probe is placed in the vagina. The saline inside the uterus gives the doctor a good view of the inside of the uterus and can be used to diagnose certain types of fibroids, polyps and endometriosis as well as other structural abnormalities. A sonohistogram is usually carried out at the doctor's office and takes between 20 and 30 minutes.

Although a sonohistogram can detect many of the same uterine abnormalities as an HSG, it cannot usually detect whether the fallopian tubes are open. For this reason, sonohistograms may be reserved for donor egg recipients or patients who do not need to have a full investigation of the fallopian tubes (e.g. women who have had previous pregnancies, tubal ligation or a recent HSG test).

HYSTEROSCOPY

A hysteroscopy is used to treat uterine abnormalities such as polyps or fibroids discovered during an HSG, or to take a closer look at other abnormal findings. It is not part of the routine testing of infertility.

During this procedure the doctor examines the uterine lining in great detail by passing a thin viewing tool through the cervix into the uterus. The hysteroscope has a light and camera attached and enables the doctor to see inside the uterus on a video screen. This technique is used to identify problems such as scar tissue in the uterus and can identify abnormalities in the shape and size of the uterus. The hysteroscope can be used to identify and remove polyps or fibroids in the uterine lining that may be interfering with fertility.

This test can be done at the doctor's office under local anesthesia or may be carried out at the hospital under general anesthesia depending on the doctor. The procedure and recovery time are usually quite short.

LAPAROSCOPY

A laparoscopy is not used as a diagnostic tool for infertility; it is very rarely done nowadays and is reserved as a diagnosis and treatment option for people with severe pelvic pain.

This is a surgical procedure to look inside the body through a small incision in the abdomen. It is a direct visualization of the ovaries, uterus and outside of the fallopian tubes. A laparoscope allows the surgeon to identify whether there is any scar tissue, endometriosis or fibroids clearly visible around the reproductive organs as well as any other abnormalities with the uterus, fallopian tubes and ovaries. Because a laparoscopy requires anesthesia and is an invasive technique, it is not used as commonly as other methods of diagnosis such as HSG, which are less complicated and carry less risks.

ENDOMETRIAL BIOPSY

An endometrial biopsy is rarely done and involves taking a small sample of the uterine lining for examination under a microscope. This test is used to determine whether the lining is fully developed and able to support a pregnancy. It can also detect infections and inflammation of the uterine lining. Endometrial biopsy is usually done three or four days before your menstrual period is due.

Another test performed on tissue taken during an endometrial biopsy is the detection of beta 3 integrins during the time at which the uterus is receptive to implantation. Beta 3 integrins are proteins, which help the embryo to implant in the uterus. Some studies have shown women with low or absent beta 3 integrins have a lower chance of getting pregnant compared with women who have high levels. If the test results from the endometrial biopsy are abnormal your doctor may prescribe additional hormonal support or a change in stimulation protocol during the treatment cycle.

BLOOD TEST FOR WOMEN WITH A HISTORY OF MISCARRIAGE

Women who have a history of two or more miscarriages may be tested for Lupus Anticoagulant (LAC), Anti cardiolipin antibody (ACL) and MTHFR mutation. These blood tests identify if there are immune or clotting disorders in the blood. Treatments may include baby aspirin, heparin, prednisone or ivIg (intravenous immunoglobulin) during the conception and early pregnancy phases.

GENETIC ANALYSIS

A full chromosome analysis may be carried out to determine your "karyotype" or genetic makeup, especially if there is a long history of unexplained infertility or recurrent miscarriages. The analysis is a simple blood test that can detect extra or missing chromosomes, deletions, additions or rearrangements within the chromosomes, such as translocations. Most people do not require a karyotype.

The following table shows the various diagnoses you can be given for infertility and how the doctor tests for each of them.

DIAGNOSES:	TESTED BY:
Ovulation disorders or hormone imbalances	Blood tests for LH, Estrogen, Thyroid Hormones, Progesterone, Prolactin etc. Ultrasound exam of ovaries can identify polycystic ovaries
Endometriosis	Medical history, symptoms, physical exam, ultrasound and laparoscopy
Blocked tubes	Hysterosalpingogram, Sonohistogram
Diminished ovarian reserve (or poor egg quality)	Blood tests for AMH, Day 3 FSH and/or a Clomiphene Challenge Test Resting (antral) follicle count by ultrasound
Male factor	Sperm count, morphology, motility, functional tests, genetic tests
Genetic disorders	Family history. Genetic tests of both partners including karyotype. Pre-implantation genetic testing of embryos
Uterine factors such as fibroids and polyps	Ultrasound, Hysterosalpingogram, Sonohistogram, Hysteroscopy
Immunity/blood clotting disorders	Blood tests, Genetic testing
Unexplained infertility	Failing to find a diagnosis

Tests for Diagnosis of Male Infertility

Semen analysis is performed early in the diagnostic phase. Almost a third of men are found to have some problems with their specimen, however, the male component does not contribute to the overall pregnancy success rate as much as the egg does. The reason for this is that women are born with all their eggs, which diminish with time and are subject to the aging process. Men continually make new sperm and so they do not have the same chromosomal changes with time that eggs do.

It takes approximately 12 weeks for sperm to be made, and so the sample produced on the day of the IUI or egg collection has been in the body for 12 weeks. It is a good idea to adopt a healthy lifestyle during the months prior to fertility treatments or conception attempts in order to maximize the chances of getting the best sperm on the day. The clinic usually tells you to have three days of abstinence prior to producing your sample. Before this abstinence period it is beneficial to ejaculate regularly in the weeks leading up to the cycle. Studies have shown that the DNA can be better in sperm samples following frequent ejaculation. Don't worry if the volume or count is lower, the quality will be better and that's what matters most.

The parameters tested in a routine semen analysis commonly include:

- Volume
- Sperm count
- Sperm motility
- Sperm morphology
- Strict morphology (Kruger)
- Antibody testing
- pH
- Presence of debris or cells
- Fructose testing

Functional tests may also be performed; these include a "swim up" which is where the sperm have to swim over a certain distance of warm media or a sperm penetration assay (SPA) where the sperm are incubated with hamster eggs to see if they bind to the shell. These functional tests are used to predict whether the sperm will be able to fertilize the eggs naturally. The semen analysis results along with the medical history of both partners will be used to make a decision about the best course of treatment.

A slide showing a sperm count. A small drop of the semen sample is placed on a glass slide and grid is used to count the sperm present. This tells us how many millions per milliliter there are.

If a man has a history of chemotherapy, hernia repair, varicocele or a vasectomy reversal, he is likely to have a low sperm count. When the reason for a low sperm count is known, further testing may not be performed. A man with an unexplained low sperm count should be referred to a urologist for an evaluation. If a physical examination reveals no abnormalities then genetic testing may be carried out.

Genetic abnormalities can manifest in low sperm counts or no sperm at all (called azoospermia). One common reason for azoopsermia is when a man carries a mutation in the gene which causes cystic fibrosis. Although the man himself does not suffer from cystic fibrosis, the presence of this gene mutation has prevented the development of the Vas Deferens, the tube that allows the sperm to leave the body. Absence of the Vas Deferens results in no sperm being present in the ejaculate. In most cases, there is sperm present in the testes that can be extracted surgically for use in an IVF cycle with ICSI.

SURGICAL EXTRACTION OF SPERM

The processes for surgically removing sperm are called Microsurgical Epididymal Sperm Aspiration (MESA), Testicular sperm extraction (TESA) or a Percutaneous Epididymal Sperm Aspiration (PESA). All involve using a needle injected directly into the testicles. Alternatively a biopsy may be taken and processed in the lab to separate out the sperm from the other cells.

A skilled urologist performs this procedure while the patient is under local or general anesthesia. Recovery times vary depending on the degree of invasion and exploration required. Following surgical removal of the sperm, an IVF cycle with ICSI is necessary to fertilize the eggs even if the yield is high.

For men who have had a vasectomy and whose partners have no known fertility problems, a vasectomy reversal is recommended over IVF (ASRM guidelines).

GENETIC TESTING

Genetic testing may include a karyotype, which detects major chromosomal abnormalities such as extra or missing chromosomes, or rearrangements of the chromosomal arms called translocations. The presence of this type of abnormality could cause recurrent miscarriage or implantation failure.

Some men with a low sperm count carry small deletions in their Y chromosome. A blood test can detect this condition although it is quite expensive and may not change the treatment process. In other words, nothing can be done to correct this condition and ICSI is usually necessary to fertilize the eggs. At worst, the Y chromosome deletion could be passed down from father to son but the children are usually healthy and normal (male offspring may have fertility problems in the future; this is currently unknown).

SPERM CHROMATIN STRUCTURE ASSAY (SCSA)

A Sperm Chromatin Structure Assay (SCSA), detects the degree of DNA fragmentation in the sperm. High levels of fragmentation can indicate a potential problem with the sperm and can identify men who have a lower chance of achieving pregnancy. DNA fragmentation in sperm may be the result of many factors including, but not limited to, oxidative stress, disease, diet, drug use, high fever, elevated testicular temperature, air pollution, cigarette smoking and age (late 40's and beyond). Although not much can be done clinically to improve the sperm, this test provides information that may help a man to adopt a healthy lifestyle and can help determine future treatment plans.

What You Can Do to Improve Sperm Quality

- Keep cool, avoid hot tubs
- Take a multivitamin that includes Vitamin C, Zinc and Folic acid
- Do not take prescription drugs unless necessary
- Reduce alcohol and eliminate cigarettes
- Eat plenty of fruits and vegetables rich in antioxidants, such as blueberries
- Reduce stress
- Exercise regularly
- Maintain a healthy weight
- Ejaculate frequently, with an abstinence of 2–3 days prior to treatment cycles

MEDICATION AND DRUGS KNOWN TO AFFECT MALE FERTILITY

Recreational drugs

- Alcohol
- Cigarettes
- Marijuana
- Opiates

Medications

- Anabolic steroids
- Antihypertensives
- Allopurinol
- Erythromycin
- Chemotherapy
- Cimetidine
- Colchicine
- Cyclosporine
- Dilantin
- Gentamycin
- Nitrofurantoin
- Tetracycline

This is not a full list of all the medications that may affect male fertility. You should inform your doctor of any prescription or over-the-counter medication you are taking.

Once the doctor has fully reviewed both partners' medical histories and performed all the necessary tests, he will recommend the course of treatment most suited to you. The next chapter will explore these options.

3

Treatment Options

Fertility treatment options include intrauterine insemination (IUI or Artificial Insemination) or In Vitro Fertilization (IVF). Depending on the diagnosis, the first line of treatment usually involves several cycles of Clomid IUI; this is especially true for young women who have a diagnosis of ovulation disorders, polycystic ovaries or couples with unexplained infertility. If a woman is not pregnant following 3–4 cycles of Clomid IUI, there is an option to proceed to more highly medicated IUI's or directly to IVF.

Some people bypass IUI's altogether and proceed directly to IVF following infertility testing, this is always the case if the fallopian tubes are blocked or the sperm count is found to be severely low. The decision of how quickly to progress to IVF is based on the age of the female partner, medical history, patient preference and financial constraints.

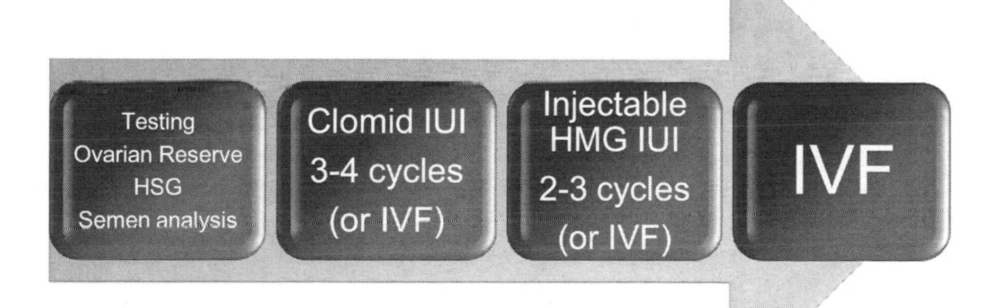

Intrauterine Insemination (IUI)

Intrauterine insemination is a procedure where the sperm is placed directly into the uterus. This is done by placing a long flexible catheter through the cervix and gently expelling the sperm. Prior to the IUI, the semen sample is collected by masturbation. Due to its alkaline pH, the seminal fluid can irritate the uterus, so the sperm is washed prior to placement by spinning it in a centrifuge.

Eggs are not retrieved for an IUI; fertilization and embryo development happen inside the body. IUI can only be used if the fallopian tubes are open and ovulation occurs. IUI has shown to be helpful in some cases of low sperm counts as the sample is centrifuged prior to insemination and the sperm are concentrated into a small volume of fluid. This allows more sperm to be introduced into the uterus than would happen following intercourse. IUI is also a good way to circumvent the cervical mucus, which may be "hostile" to sperm and prevent it from swimming through the cervix.

The IUI should be timed to co-ordinate with ovulation so you may be asked to monitor your cycle with home ovulation predictor kits. When you see a positive surge in Luteinizing Hormone (the hormone which triggers ovulation) you call the clinic and make an appointment for your insemination the next day. Intercourse (and orgasm!) are encouraged following IUI in order to get the maximum amount of sperm to the egg as possible. Check first with your doctor to make sure there was no bleeding during the IUI, which is very rare.

Success rates of IUI vary considerably and are based on the number of follicles present, the sperm count and any other underlying factors that are preventing you from getting pregnant.

NATURAL IUI CYCLE

A natural cycle IUI is usually recommended for women using donor sperm who have no known fertility problems. The success rates for natural IUI cycles are not much higher than natural conception for people with unexplained infertility. The overall success rate per cycle is somewhere around 10% and the cost is $200–$300. Depending on your age and diagnosis most doctors would recommend doing no more than 3–4 cycles of natural IUI before moving on to medicated IUI and/or IVF.

MEDICATED IUI CYCLE

A medicated IUI cycle follows the same basic principles as a natural cycle IUI. Both involve washing and concentrating the sperm and placement of the sperm into the uterus. With a medicated IUI cycle you will take drugs to promote follicular development and ovulation.

The success rate of medicated IUI is approximately 20% per cycle, slightly higher than natural cycle IUI's. The downside is the increased risk of a multiple pregnancy, especially when using injectable gonadotrophins. For that reason, it is important to have regular ultrasound monitoring of the ovaries leading up to the insemination. If many follicles develop, you may be advised to cancel the cycle and use a lower dose next time. Alternatively, an IUI cycle can be converted into an IVF cycle if there is an unexpected high response.

One drug regularly used in medicated IUI cycles is Clomid, also known as Clomiphene Citrate. Clomid works on the brain and promotes the production of Follicle Stimulating Hormone (FSH). This hormone stimulates the growth of follicles in the ovary and is a first line of treatment to induce ovulation. The use of Clomid has a 5% chance of twins and a very rare chance of triplets. Because Clomid can change the cervical mucus and make it difficult for the sperm to penetrate into the uterus and fallopian tubes, it is often recommended that Clomid be used in conjunction with intrauterine insemination, rather than intercourse, to maximize the chance of pregnancy. Clomid is relatively inexpensive costing only a few dollars. However, a Clomid IUI cycle could incur additional expense because of extra ultrasound monitoring and blood testing.

Letrozole is a drug that has similar affects to Clomid and similar success rates. It is possible that patients who don't respond well to Clomid, or have intolerable side effects, may do better using Letrozole.

If Clomid or Letrozole IUI is not successful you may be advised to advance to the next level of treatment. This could be IUI with injectable gonadotrophins or IVF. It is not advised to do more than 6 Clomid IUI cycles.

Injectable gonadotrophins are the same drugs used for IVF to produce many eggs at once. For an IUI cycle, the dose of gonadotrophins is much lower than an IVF cycle with the intention of developing only 2 or 3 follicles before insemination. Because people have different responses to these injectable medications, the doctor will follow the development of the follicles with ultrasound and possibly blood tests. It is important not to inseminate or have intercourse if you are at risk of a high order multiple pregnancy (usually three or more mature follicles). Based on your age and diagnosis the doctor will assess the risks of this happening, versus the chance of achieving a pregnancy, and proceed accordingly. The cost of a gonadotrophin IUI cycle can be around $2000 per cycle.

The next level of treatment and the ultimate in technology is IVF.

In Vitro Fertilization (IVF)

During an IVF cycle, relatively high doses of injectable gonadotrophins are given to stimulate the production of many eggs. The eggs are surgically removed from the ovaries and fertilized in the laboratory. The resulting embryos are grown in a Petri dish inside a warm incubator for up to 6 days before being transferred back into the body. There is no other method to assess early embryo development than IVF and it serves as a diagnostic tool as well as a treatment option.

IVF is generally used in couples who have failed to conceive after trying to get pregnant for at least one year or who have any of the following:

1. Blocked fallopian tubes
2. Severe male factor infertility
3. Failed multiple cycles of ovarian stimulation with IUI
4. Advanced maternal age and over 6 months of infertility
5. Reduced ovarian reserve
6. Severe endometriosis

Tests and surgeries for infertility do not usually help a woman get pregnant and so it may be advantageous to start treatment rather than undergoing extensive investigation. Couples don't want to have extended and expensive testing that doesn't change the ultimate treatment plan and they don't want to wait very long to get pregnant. Treatment with Clomid IUI can start immediately, especially if the doctor is doing a clomiphene challenge test as an early diagnosis, this test can be done in conjunction with an IUI.

When evaluating whether to move onto IVF from IUI's, or bypass IUI's altogether, consider the cost effectiveness of infertility treatments. Given the higher chance of success from IVF compared with other treatments, this will be the most cost effective therapy for some couples.

IVF is the most high tech treatment available and has the highest success rates. This type of treatment will be covered in great detail for the remainder of this book.

Medication Made Simple

Whether you are doing an IUI cycle or an IVF cycle it is quite likely that you will be taking some new medication, some medicines are in tablet form and others are injected. You will probably take a combination of some of the drugs shown in the following table. It is important that you take the medication on the right day at the right time. The timing of some is quite essential and so it is best to be organized and plan ahead if you need to take your medication while you are at work or traveling.

The different medications can feel overwhelming at first, so it is a good idea to keep a calendar with all the drugs to be taken each day clearly written down. You can ask your clinic to help you with this.

If the following information about medication and stimulation protocols is more detail than you need…feel free to skip this section. You don't need to understand how the medication is working or why you are taking it; it will still do the job! The most important thing is to follow your doctor's instructions and take the correct dose on the right day and at the right time.

MEDICATIONS USED FOR FERTILITY TREATMENTS

Name of drug	What does it do?	How do you take it?	Why do you take it for fertility?
Antibiotics	Prevents infection	By mouth	Given to prevent bacterial infection following egg retrieval
Baby Aspirin	Thins the blood	By mouth	Increases blood flow to the uterus, reduces risks of blood clotting
Bromocriptine	Reduce prolactin levels	By mouth	Elevated levels of prolactin can interfere with ovulation
Clomiphene Citrate e.g. Clomid	Induces ovulation in some people by blocking estrogen receptors	By mouth	To induce ovulation and can be good for people with polycystic ovary syndrome
Corticosteroid e.g. Medrol, Prednisone, Dexamethasone	Anti-inflammatory	By mouth	Suppresses the maternal immune response to the embryo
Dehydro-epiandroterone (DHEA)	Mild male hormone	By mouth	May increase pregnancy rates in women with premature ovarian failure or diminished ovarian reserve
Estrogen e.g. Estrace, Estraderm	Naturally occurring female sex hormone	By mouth, skin patch or vaginal suppository	Stimulates the uterine lining to grow in preparation for pregnancy
Follicle Stimulating Hormone (FSH) e.g. Gonal F, Follistim, Menopur, Bravelle, Repronex	Naturally occurring in the body, FSH stimulates the ovary to produce follicles	Injection	Given to stimulate the ovaries to produce many follicles (and therefore eggs) in one cycle
Gonadotropin Releasing Hormone Agonist (GnRH-a) e.g. Lupron	Binds to receptors in the brain and reduces natural hormone levels (following an initial surge)	Injection	Prevents premature ovulation (blocks LH surge during stimulation phase of IVF cycle) Down regulates the natural hormones for a more controlled treatment cycle
Gonadotropin Releasing Hormone Antagonist e.g. Ganirelix, Cetrotide	Binds to and blocks receptors in the brain and reduces natural hormone levels	Subcutaneous injection	Prevents premature ovulation during the stimulation phase of the IVF cycle (blocks LH surge in response to rising estrogen levels)
Growth Hormone e.g. Saizen	Naturally occurring hormone	By injection (along with stimulation meds)	May increase IVF success rates in sub-optimal responders

MEDICATIONS USED FOR FERTILITY TREATMENTS *continued*

Name of drug	What does it do?	How do you take it?	Why do you take it for fertility?
Human Chorionic Gonadotrophin (HCG) e.g. Novarel, Pregnyl, Ovidrel, Profasi	Naturally occurring hormone usually released during pregnancy	Intramuscular injection	Given when the follicles are considered to be mature, hCG triggers the brain to start the ovulation process
IvIG	A protein derived from blood, has anti immunological properties	Intravenous	Suppresses the maternal immune response to the embryo
Heparin e.g. Lovenox	Anti-coagulant	Injection	Thins the blood and reduces the risk of blood clotting
Letrozole e.g. Femara	Induces ovulation in some people by blocking estrogen receptors	By mouth	Used for ovarian stimulation and to promote ovulation. Has less side effects than Clomid
Metformin	Helps control blood sugar levels	By mouth	Reduces insulin resistance in women with polycystic ovaries. Pre-treatment may improve IVF pregnancy rates
Methotrexate	Inhibits the metabolism of folic acid	By mouth or injection	Used to treat ectopic pregnancies which are dangerous pregnancies that grow outside the uterus
Prenatal Vitamins	Good general health	By mouth	Folic acid reduces the risk of certain spinal cord defects in babies
Progesterone	Naturally occurring female sex hormone	Intramuscular injection or vaginal suppository	Enhances and maintains the uterine lining, helps to support a growing fetus
Viagra	A drug to increase blood flow	Taken vaginally by women; orally by men	May increase uterine blood flow and help women with poor endometrial development

ABOUT GONADOTROPHINS (OVARIAN STIMULATION MEDICATION)

There are two types of gonadotrophins available. One is called human menopausal gonadotrophin (hMG), which is derived from the urine of postmenopausal woman. The other type is a pure form made by genetic engineering (these are called recombinant drugs).

The human derived gonadotrophins can contain some "impurities" most notably luteinizing hormone (LH). Many doctors believe that a small quantity of LH during ovarian stimulation produces a better result in some patients. For this reason it is quite common to use a mixture of human derived gonadotrophins and recombinant gonadotrophins during the stimulation phase.

Gonadotrophin Brand Names

- **Human menopausal gonadotropins (hMG)** are a mixture of LH and FSH. The names of the most widely used hMG medications are **Pergonal, Repronex and Menogon**. Some highly purified human derived gonadotrophins are called **Bravelle** and **Menopur**.
- Recombinant FSH (pure FSH) is sold under the brand names Fertinex, Follistim, Gonal-F and Puregon.

Stimulation Protocols

There are several different ways your doctor can choose to stimulate your ovaries for IVF treatment. These are called **PROTOCOLS**. Protocols vary by using different combinations and timing of drugs

The most common type of stimulation protocol is the *LONG LUPRON PROTOCOL.* The *SHORT LUPRON PROTOCOL* (also called a **micro-dose** or a **flare protocol**), and the *ANTAGONIST* protocol are also outlines below.

IMPORTANT

These are examples only. Always follow your own doctor's instructions for your cycle.

LONG LUPRON PROTOCOL (STANDARD OR OVERLAP PROTOCOL)

This type of cycle usually starts with birth control pills, which are used to regulate the body's hormones and synchronize the timing of the other drugs. Another advantage of using birth control pills to overlap the Lupron is that the pills can suppress the formation of ovarian cysts that sometimes form when Lupron is taken.

On day 21 of your cycle (or after 21 days of active pills) you begin Lupron. This is a drug that acts on the parts of the brain that communicate with the ovaries. The body's initial response to Lupron is to produce a surge of hormones that stimulate the ovary. The hormones are Follicle Stimulating Hormone and Luteinizing Hormone. This is followed by *suppression* of the ovary with continuing Lupron.

Lupron ultimately blocks the pathway between the ovaries and the control centers in the brain giving the doctors more control over the treatment cycle. It takes around 10 days of Lupron to fully reduce the body's natural hormones to a very low level. When the hormone levels are sufficiently low you are said to be "suppressed" or "down regulated." This suppression helps the ovaries to recruit multiple follicles instead of the single follicle, which would usually be recruited during a natural cycle.

The birth control pill is eventually discontinued and Lupron continued during the stimulation phase of the cycle. This phase involves the injection of drugs that stimulate the follicles to grow in the ovaries. During this time you will have various ultrasounds and blood tests to monitor the developing follicles and the dose of drugs can be altered accordingly.

Because the growing follicles produce and secrete estrogen into the blood stream, the circulating levels of estrogen are monitored closely during the stimulation phase. This level, coupled with ultrasound counts and measurements of the follicles, allows the doctors to accurately measure the body's response to the drugs.

After approximately 10 days of stimulation, when the biggest follicles are around 18–22 mm in diameter, you will be told to discontinue the stimulation medication and Lupron.

At this point you will take an hCG shot (Human Chorionic Gonadotrophin), which matures the eggs and releases them from the walls of the follicles in preparation for the egg retrieval. (hCG acts in exactly the same way as Luteinizing Hormone).

One to two days following the egg retrieval you will begin progesterone supplementation which is very important for maintaining the lining of the uterus and keeping open a window of receptivity to embryo implantation. Estrogen is often taken following the egg retrieval to maintain the right balance of hormones in the body because egg collection can remove estrogen-producing cells that are important to the process.

Long Lupron Protocol Summary

1. Birth control pills (overlap with Lupron)
2. Lupron (overlaps with stimulation medication)
3. Suppression check
4. Stimulation medication (frequent ultrasound and blood tests for approximately 10 days)
5. HCG
6. Egg Retrieval
7. Progesterone (and estrogen) supplementation
8. Embryo Transfer
9. Continued progesterone supplementation

SHORT LUPRON PROTOCOL (MICRO-DOSE OR FLARE PROTOCOL)

The short protocol takes advantage of the fact that the body's initial response to Lupron is a *surge* in hormones for several days that stimulates the ovaries. During a short protocol the Lupron is first introduced during the stimulation phase. This gives the ovaries an extra dose of natural FSH on top of the high doses being injected. A short protocol is used when people don't respond well to the long protocol. It is one of the most potent protocols available and can help women with low ovarian reserves to get as many eggs as possible.

Short Protocol Summary

1. Birth control pills
2. Mini doses of Lupron are started 3 days after the last pill and continued until the day of HCG. Continued Lupron prevents premature ovulation.
3. Ovarian Stimulation, started shortly after Lupron and overlaps it
4. HCG
5. Egg Retrieval
6. Progesterone (and estrogen) supplementation
7. Embryo Transfer
8. Continued progesterone supplementation

ANTAGONIST PROTOCOL (CETROTIDE/GANIRELIX)

The antagonist protocol uses a drug that binds to the same receptors in the brain as Lupron. The difference between the two drugs is that Lupron elicits a response upon binding and the antagonist binds to the receptors and has no effect.

In other words the agonist (Lupron) produces a response similar to the natural hormone, whereas the antagonist not only produces no response, it also blocks those receptors by occupying them.

The antagonist does not trigger the release of FSH and LH like Lupron does. Its action is very similar to Lupron in its suppressive phase and thereby **prevents premature ovulation**. The antagonist is taken during the stimulation phase of the treatment cycle when the follicles are around 14mm in diameter and is continued until the HCG shot.

An advantage of the antagonist protocol is that the drug only takes one day to suppress the natural hormones that may interfere with the cycle. In addition there are fewer injections and there may be a better response in people with a low ovarian reserve due to the fact that the ovaries are not subjected to prolonged suppression prior to the stimulation phase.

Antagonist Protocol Summary

1. Birth control pills
2. Ovarian stimulation is started at the end of the pill cycle
3. Antagonist (Cetrotide or Ganirelix) is started 4 to 5 days after the stimulation medication, continued until the day of HCG
4. HCG
5. Egg retrieval
6. Progesterone (and estrogen) supplementation
7. Embryo transfer
8. Continued progesterone supplementation

As you can see there are several different options available for stimulating the ovaries and sometimes the first attempt doesn't get the best response. In this case it might be necessary to cancel a cycle after you have started the medications and restart on a different protocol. Although this is disappointing, it is better to cut your losses and start over with a better understanding of your body.

Donors and Carriers

USING DONOR EGGS

Donor egg IVF is an excellent option for women with a very low chance of conceiving with their own eggs. The success rates at the top clinics in the USA are close to 80% live births and so the chance of having a baby is very high. The woman has a close bonding experience by carrying the baby and giving birth, as well as having control over the baby's health during the gestation period. Women who have children using donor eggs have reported that they feel like the baby is truly "theirs" and couples are delighted that they chose donor eggs as part of their journey to have a family.

The doctor may recommend donor eggs when a woman has a low ovarian reserve, is of advanced age or carries a hereditary disease. There are always clinics who will try IVF with a woman's own eggs in spite of low odds and this may be something that is important for a couple to pursue before making a decision to use donor eggs.

As a couple, you should be fully informed about your chances of success and the cost of treatment both financially and emotionally, before you decide what option is right for you. Often times the donor egg IVF experience is a welcome change from previous frustrations and disappointments with treatments using your own eggs.

Choosing donor eggs can, however, be a difficult decision for some women who feel a profound loss over the genetic link to the child; these feelings are normal. Once a choice has been made to pursue this option it is very important to have support and guidance through the process. Most clinics advise a therapy session for the couple to ensure that using donor eggs is the right decision for their family.

The most important factor when starting a donor egg cycle is to choose a good clinic. Check the SART website and chat with people on www.ivfconnections.com or www.pved.com for advice. Once you choose a clinic they will guide you in the process of choosing a donor and where to go from there. Some clinics show photographs of the donors and all clinics will give you a full profile detailing physical characteristics as well as medical and family histories. Do not be afraid to travel out of your town for an egg donor IVF cycle. If you find a donor and a clinic you feel comfortable with you should invest a little more to get the best match and service possible. The overall cost of an egg donor cycle can be in excess of $25,000. This includes donor fees and the costs incurred from the IVF cycle itself. The cost itself is one really good reason to choose the very best clinic when starting a donor egg cycle.

Your doctor will conduct several tests to ensure your uterus is normal and your overall health is good. A consultation with a high-risk doctor (perinatologist) may be recommended to women over 45 years old. Most egg donor cycles use anonymous donors but some people choose to use a sister, a family member or a good friend. It is important for everyone involved to talk with a mental health professional to explore the feelings that may arise as a result of known donation.

In the USA there are a handful of frozen donor egg "banks" where you can buy several eggs to be shipped to an IVF clinic for treatment. This may seem like a more cost effective treatment option but the overall success rates are lower than with fresh eggs and embryos. An advantage egg banks do have is that the eggs are quarantined before use. With a fresh donor egg cycle there is a very small chance of contracting a communicable disease although the donors themselves are thoroughly tested before and during the donation process.

IVF Donor Egg Cycle Process

The egg donor will go through a stimulated IVF cycle, the same way as a woman using her own eggs would. At the same time, the egg recipient will take medication to synchronize her cycle with the donor. The recipient's uterine lining has to be at the right stage of receptivity at the time the embryos are ready for transfer. The IVF clinic will facilitate this synchronization as well as keep you updated on the donor's progress during the stimulation phase. The husband or male partner will be expected to give a sperm sample on the day of the egg retrieval. Once the eggs are retrieved, they are fertilized by standard insemination or ICSI and the IVF cycle progresses in the same way as any other IVF cycle; embryos are grown in the lab until the day of transfer. The best ones transferred into the uterus of the recipient and excess embryos of good quality are cryopreserved for future use. Pregnancy tests are usually done 14 days after egg retrieval.

USING DONOR SPERM

Donor sperm is readily available from several sperm banks and some IVF clinics have their own "in house" donor sperm ready for use. Most sperm banks offer a full profile of the donors including physical characteristics, medical background and educational credentials. You may also be able to see current or childhood photographs of the donor.

Sperm donors are carefully selected to ensure normal fertility and a good genetic and medical background. As with donor egg recipients, people who choose to use donor sperm should consult with a therapist to make sure this is the right decision for them. Most sperm donation is anonymous with little chance of future contact with the donor, although some people choose to use a known donor. In this case, carefully drawn legal documents should be in place prior to treatment outlining the legal rights of the donor should a conception occur.

Donor sperm is commonly used with artificial insemination or IUI, especially if the woman does not have any fertility problems herself. Sperm from a donor can also be used for IVF. Some clinics routinely perform ICSI when using donor sperm, because the functionality of the sperm can be reduced when it is frozen and thawed.

All sperm donors are thoroughly tested for communicable and genetic diseases at the time of donation. The sperm is then quarantined for 6 months at which point the donor is re-tested. If all the test results are clear only then are the samples released for use.

DONOR EMBRYOS

Embryos frozen during an IVF cycle can be donated to another couple if the original parents have completed their family. Several agencies facilitate this process, which is sometimes referred to as embryo adoption.

The donating couple must undergo psychological counseling and communicable disease testing in order for the embryos to become available. Embryo adoption is an excellent option for couples that are unsuccessful in achieving a pregnancy or have suffered multiple pregnancy losses. Many couples consider donor embryos a very early adoption. Couples who donate their embryos want to help other people who are going through the same frustrations trying to conceive that they also went through.

The pregnancy rates from donor embryos are dependent on the age of the female partner at the time of freezing as well as the number available and the stage at which the embryos are frozen. If you are interested in embryo adoption, first ask your clinic if they have any embryos available. This is a less expensive and less time consuming option than using an agency.

For more information, contact Snowflake Embryo Adoption Agency, The National Embryo Donation Center, Reprotech or Miracles Waiting (as well as other agencies that can be found on the Internet).

USING A GESTATIONAL CARRIER

Using a gestational carrier (GC) involves having another woman carry and deliver your baby for you. In most situations a couple will undergo a cycle of IVF with the embryos being placed in the uterus of the gestational carrier, who will carry the baby to term. When the child is born, the carrier will sign away all parental rights and give the baby to the biological parents.

Several agencies specialize in matching carriers and couples and coordinate all the necessary legal and medical requirements. Success rates depend on many factors mostly the age of the woman who donates the eggs and any underlying fertility problems that the couple may have.

It is vitally important that the relationship between the GC and the intended parents is based on trust, and some couples prefer to use a family member or friend they know very well. The legal concerns combined with intensive fertility treatments can make this a very stressful process to go through. If using a GC through an agency the cost can be upwards of $40,000.

Legal Considerations

No matter what fertility therapy you and your spouse take advantage of, it is important to be aware of the basic legal implications of the process. Remember this book is not intended to be legal advice and because the laws applying to alternative reproductive therapy and the resulting rights of individuals are not geographically uniform, you are encouraged to seek individual legal counsel. In any reproductive therapy involving the help of a gestational carrier or utilizing the genetic material of a third party it is important to have in place a written agreement with the intended parents, the donor, and/or the gestational carrier (and also with the spouse of the carrier).

The agreement should address issues relating insurance for the donor/gestational carrier, issues related to expected procedures for the gestational process, issues related to the expected rights or waiver of rights for each party, as well as other important matters. Because the law of the location where the contract is signed usually applies to a contract, it is important you know the law of the location/jurisdiction you are in before signing an agreement setting individual rights and responsibilities. It is also important to have all parties sign an informed consent articulating the intent of the agreement and ultimately confirming whether each party intends to maintain a legal relationship with the intended child or intends to waive that right.

It is best to consult an attorney familiar with the laws applying to alternative reproductive therapy issues in your location. In some instances and depending on your location you may have the option of obtaining a court order confirming parentage prior to birth. If so this court order can reduce uncertainties about the child's legal status before the birth and therefore make the experience in the hospital more enjoyable. Consulting an attorney in the early stages of your family planning can add a level of certainty that is invaluable to your family planning.

4

IVF: From Egg Collection to Transfer

So far we have looked at choosing a clinic, getting a diagnosis and how to stimulate the ovaries in preparation for an IVF cycle. This chapter will cover what happens after your eggs are retrieved, from fertilization until embryo transfer.

Inside the IVF Laboratory

The embryology lab is a very special place where the sperm will meet the eggs for fertilization and the resulting embryos will be grown for up to six days before the chosen embryos are placed back with the mother.

All the equipment in the laboratory is specially chosen to provide the perfect environment for the embryos to have their nutritional requirements met while being protected from the outside world. Incubators are kept at body temperature and special gases are introduced to create an atmosphere that closely resembles the natural conditions of the fallopian tubes and uterus.

The embryos are grown in culture media designed to mimic the fluid in the fallopian tubes for the first three days and then the embryos are transferred to different culture media, which is formulated to be more like the uterine environment. Much research has been done to determine which nutrients and growth factors are needed for optimal growth of the embryos outside the body and the culture media is designed accordingly.

Once the female partner has taken the stimulation medication for about 10 days she will be ready to take a trigger shot (hCG shot, which acts the same way as LH) in preparation for the egg retrieval two days later. The trigger shot is given when the follicles in the ovary reach the appropriate size, usually between 18–22mm in diameter. The timing of this injection is very important, as it will ensure that the eggs have matured as much as possible by the time of the retrieval.

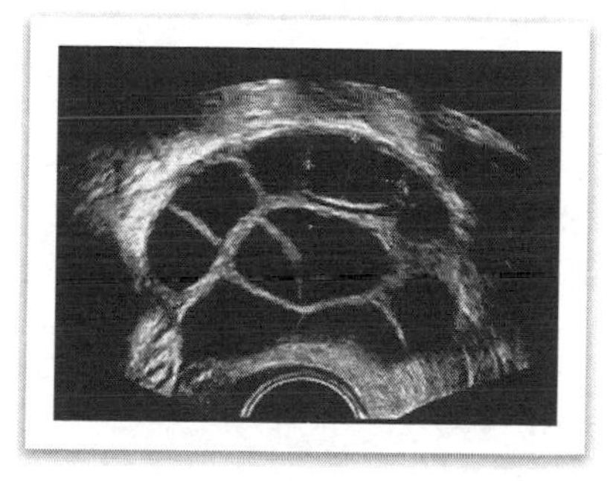

An ultrasound picture of a stimulated ovary, the follicles appear as black circles

EGG RETRIEVAL

Most egg retrievals are done under light anesthesia and are painless; however, some people do feel discomfort for the remainder of that day. To retrieve the eggs a long hollow needle is inserted into the ovary through the wall of the vagina under ultrasound guidance. An IVF doctor and a team of nurses perform this procedure. When the needle punctures a follicle, suction is applied and the fluid containing the egg is drained from the follicle and handed to the embryologist. The embryologist then identifies and isolates the eggs from the fluid.

During the egg collection the male partner usually stays in the recovery area or produces his sperm sample. After the retrieval the woman should take it easy and not go to work or do anything strenuous.

Eggs freshly retrieved from the ovary are surrounded by lots of cells called cumulus cells

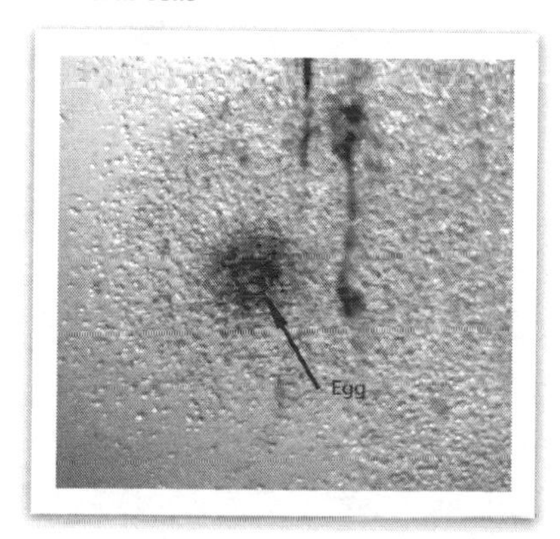

It is important to remember a few things about the egg retrieval. Firstly, the number of follicles seen on ultrasound is an *approximate* number of eggs you can expect to get; sometimes there are more, sometimes less. Secondly, not all the eggs will be mature, and the immature eggs cannot fertilize and will not be useable. The following photographs show the various stages of egg maturity. Only the eggs in metaphase 2 are mature, this can be seen by the presence of a tiny circle at the edge of the egg called the POLAR BODY. If doing ICSI, only the mature eggs are injected.

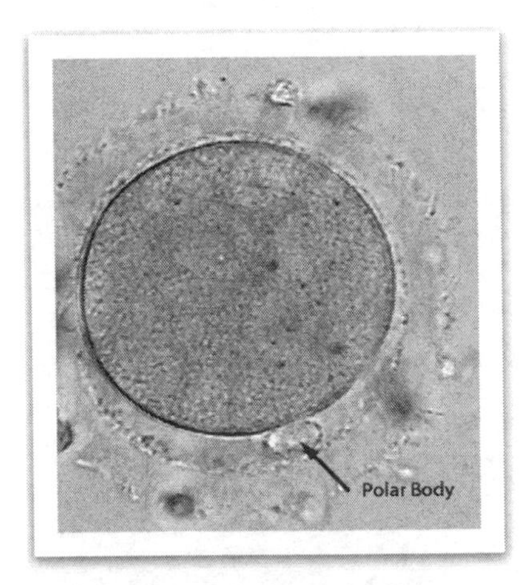

METAPHASE 2

A mature egg; polar body can be seen

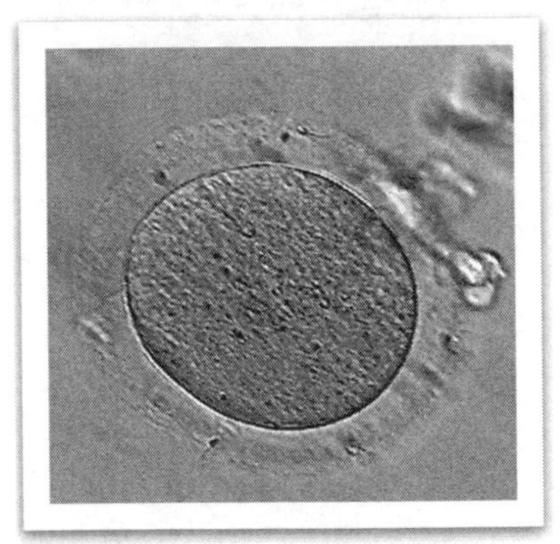

METAPHASE 1

An immature egg; no polar body is present

On the day of the egg retrieval very little information about the quality of the eggs is available. At most, we can tell you the number of eggs and possibly how many are mature. Most eggs look identical at this stage and the difference in quality only becomes evident after one to two days when the embryos have started to divide and grow.

If the husband or partner is giving a fresh semen sample that day, which is usually the case, (alternatively a sample may be previously frozen), he will be asked to do so sometime around the time of egg retrieval. The lab will prepare the sperm by separating the sperm from the seminal fluid. In most cases, it is spun in a centrifuge using a gradient of a solution that allows the separation to occur. The eggs and sperm will be kept warm for the rest of the day until it's time to inseminate the eggs.

A WORD ABOUT ABSTINENCE

The husband or partner will be asked to abstain from ejaculating for approximately two to three days prior to the egg retrieval. Recent studies have shown that frequent ejaculation can improve sperm quality and reduce DNA fragmentation that could lead to poor quality embryos. The male should abstain for two or three days but ejaculate frequently in the weeks leading up to the IVF cycle.

From egg retrieval to embryo transfer we see a dramatic reduction in the number of viable eggs and embryos. This is one of the reasons we stimulate the ovaries to produce many eggs at once, as there is a better chance of having good quality embryos for replacement back into the body the more embryos we have to choose from.

From follicle count to good quality blastocysts, the number of viable eggs and embryos decline. The most important thing is to have good quality embryos for transfer.

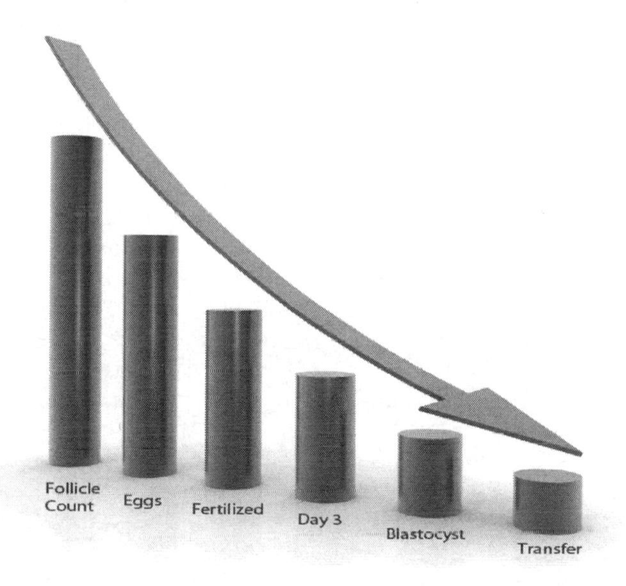

ICSI or Insemination? How We Fertilize the Eggs

If there is any concern about the sperms ability to fertilize the eggs, the embryologist may decide to perform ICSI (intra-cytoplasmic sperm injection).

Simply put, a single, normal appearing sperm is selected using a high power microscope and directly injected into an egg. This is a very effective way to fertilize the eggs and ensures (to a degree) that the eggs will fertilize.

This procedure is used in cases when the male partner has a low sperm count, low motility or morphology, surgically extracted sperm or a vasectomy reversal. Some clinics exclusively do ICSI on all their patients to avoid the rare case when there is a total fertilization failure after insemination.

Some clinics may do ICSI when there is unexplained infertility and are concerned whether the egg and sperm can bind together. Also, it is possible to do a "partial" ICSI if there is some doubt about whether the eggs will fertilize naturally. In this situation, some eggs would be mixed with sperm and some would be injected. The idea being that if the sperm don't fertilize the eggs naturally, the ICSI injected eggs are the "insurance."

On the day of the egg retrieval, the eggs and sperm will be added together by placing several thousand sperm around the eggs and allowing them to fertilize "naturally" or by doing ICSI. They will be left overnight and examined the next morning for signs of fertilization.

ICSI or insemination is performed to fertilize the eggs on the day of the egg retrieval.

ICSI- Intra-cytoplasmic sperm injection. Sperm inside needle prior to injection

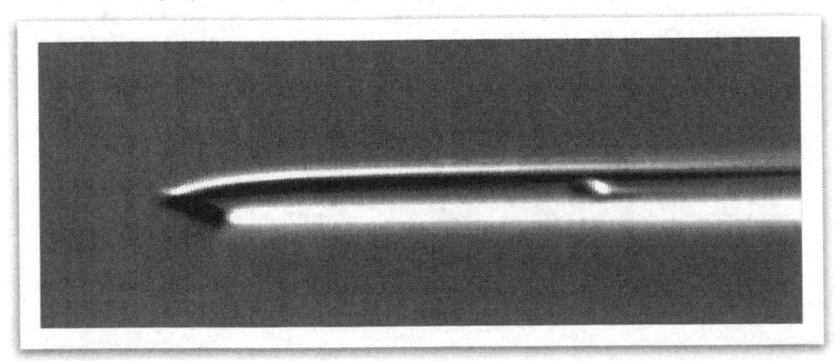

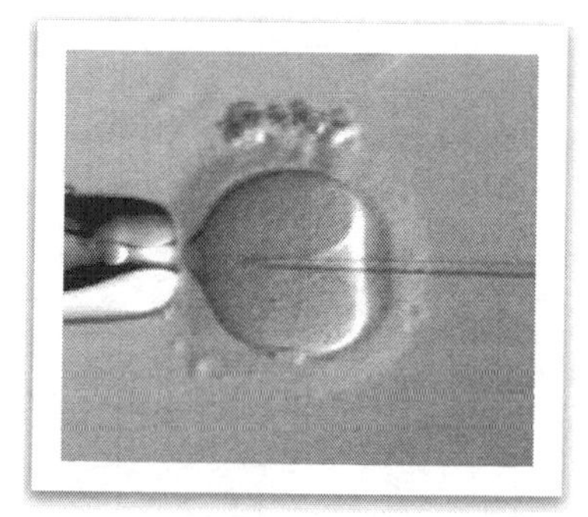

A single sperm is injected into the egg

IS ICSI SAFE?

Many studies have followed the development of children born using ICSI to ensure the safety of this technique. Several major studies have shown that although fertility treatments themselves do carry small risks, there is no difference in the occurrence of abnormalities between babies born from ICSI or standard IVF where the eggs were not injected

It is not yet clear whether genetic abnormalities are being passed from father to son through ICSI. For example, if the male partner has a chromosomal abnormality that is the *cause* of his low sperm count; this could be passed to the male children. Although the babies appear to be healthy, they may have inadvertently inherited a fertility problem themselves. This is yet to be seen as the first ICSI babies were born after 1992 and have yet to be studied into adulthood.

Stages of Embryonic Development

FERTILIZATION–DAY 1

The first thing your embryologist checks for on the morning after the egg retrieval is fertilization. The day of fertilization is counted as day 1 of development.

Fertilization is evident by the appearance of 2 circles inside the egg called PRONUCLEI. One circle contains the genetic material from the sperm and one contains the genetic material from the egg. These pronuclei fuse together within a few hours. At this stage of development the embryo is referred to as a ZYGOTE.

Sometimes more than one sperm enters the egg and there are three or more pronuclei in a single egg. These are carefully isolated from the other fertilized eggs as they do not implant and grow normally. On average approximately 80% of the **mature** eggs will fertilize normally.

Normally fertilized eggs with two pronuclei (Day 1 of development)

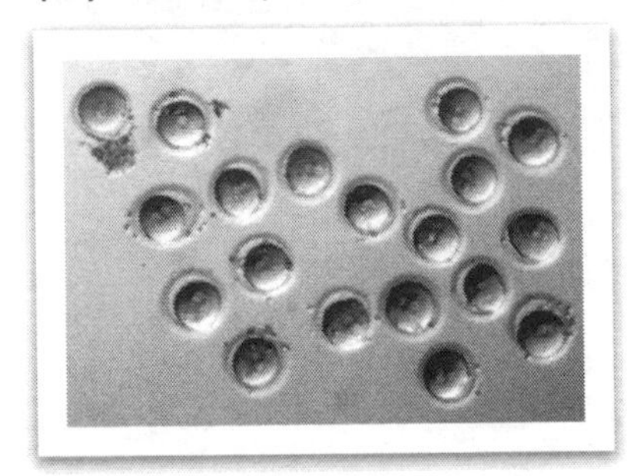

The day after the egg retrieval (day one) is very important, as you will know how many embryos you have. The lab staff or one of the nurses will call you with this information. Most patients are curious about the quality of the embryos. At this stage the embryos look identical and it is not yet possible to distinguish between one embryo and another. The embryologist will likely not have much additional information to share besides the number fertilized.

CLEAVAGE—DAY 2 AND 3 OF DEVELOPMENT

On day two of development we expect the embryos to be at the two or four cell stage and on day three, they should be six or eight cells.

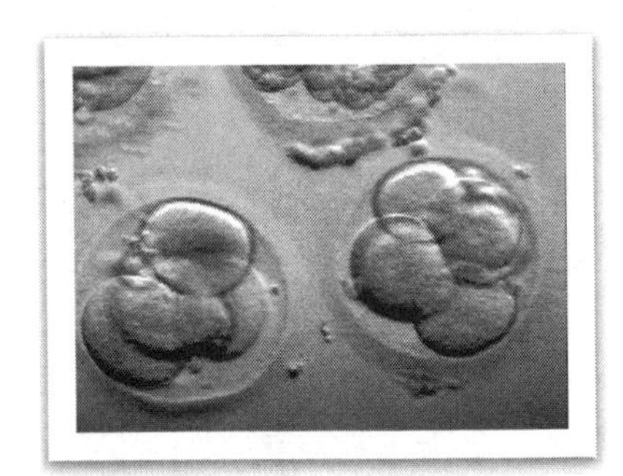

Day Two embryos- Four cells

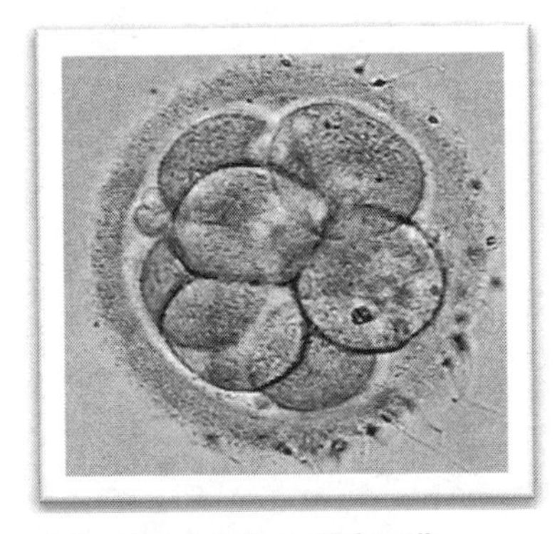

A Day Three embryo- Eight cells

Once the embryos start to divide it is possible to grade their quality. Each clinic will have its own grading system, so you will need to ask what this is to fully understand the information.

Embryo Grading

Embryo grading is generally based on:

- The number of cells—should be appropriate for the day of development
- Size, shape and degree of diversity in the size of cells—an even number of equally sized cells is best
- Clarity of the cytoplasm inside the cells—the interior of the cells should be free of inclusions and dark areas
- Extent of fragmentation—fragmentation occurs when a cell divides and parts of the cell break off. This causes the embryo to have smaller buds or fragments of cells surrounding the larger cells.

For example, a lab might have a grading system from 1 to 4, where 1 is the best quality and 4 is the poorest quality.

For cleavage stage embryos (days two and three of development) the grade will include 2 numbers; the first is the number of cells and the second reflects the overall quality of the embryo. An 8/1 would be an eight-celled embryo of excellent quality; a 6/3 would be a six-celled embryo of poor quality and so on.

You can see from the following photographs that the grade 3 embryo has many "fragments" or small pieces of cell that have broken away from the larger cells during division. Although these embryos have a lower chance of implanting than their more handsome counterparts, it is still possible to achieve a pregnancy from embryos which are not given perfect grades.

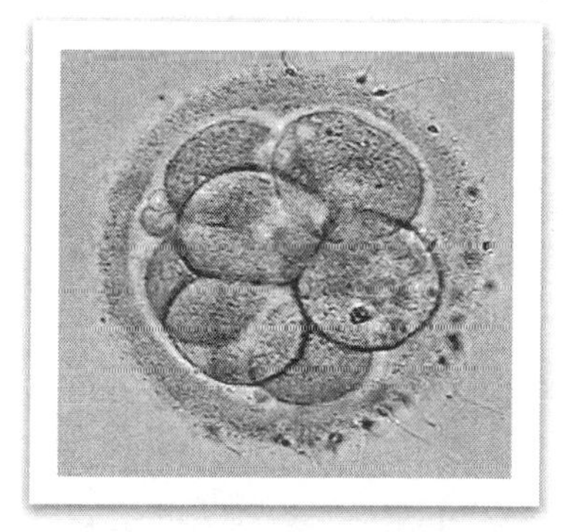

A day 3 embryo of excellent quality
Grade 8/1 (Eight cell, grade one)

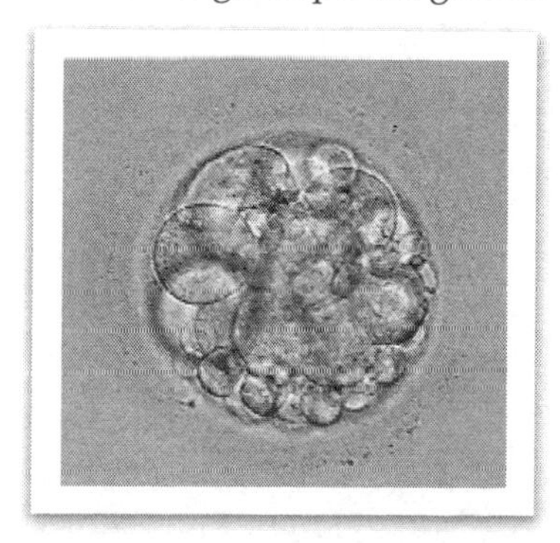

A day 3 embryo of poor quality
Grade 8/3 (Eight cell, grade three)

In the lab we usually see a wide range of quality within a "batch" of embryos. It is rare to see all embryos looking identical once they start to divide. This can be seen in the following photograph which shows the diversity in quality within a group of embryos on day 3 of development.

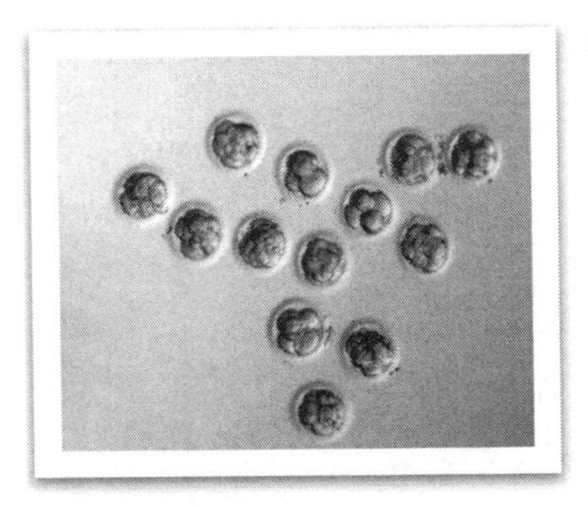

Embryos on day 3 of development

TRANSITION–DAY 4

Following the 8-cell stage, the cells of the embryo merge together to form what is called a MORULA. This is a Latin term derived from the word for mulberry. Down the microscope the embryo almost looks like a single cell again. In order for the cells to merge they have to express the correct molecules on the surface of the cells. Good quality embryos have a better ability to do this.

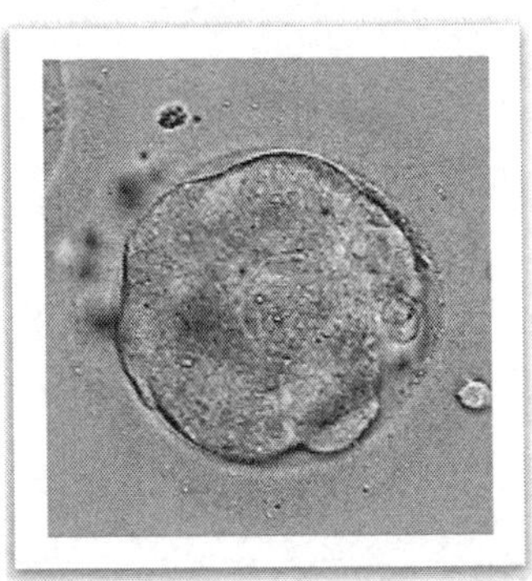

An embryo at the morula stage of development (day 4)

Between days three and day five of development, many important changes happen. This is the stage when an embryo is most likely to stop growing or arrest in its development. One of the reasons for this is that after day 3 the male chromosomes begin to contribute to the development of the embryo. Up until this stage, the cell divisions have been powered by only the egg's energy. Any abnormalities from the sperm may slow down or stop the embryos from growing past day 3.

BLASTOCYST DEVELOPMENT—DAY 5 AND 6

Following the morula stage is the all-important BLASTOCYST stage of development. This is when the embryo takes in fluid to form a cavity and the cells begin to differentiate into two different types. These two types of cells are called the TROPHECTODERM (T) and the INNER CELL MASS (ICM).

The trophectoderm cells are a single layer of cells around the circumference of the embryo which gives rise to the placenta and embryonic sac. The inner cell mass is a distinct clump of cells which will form the actual baby.

A blastocyst stage embryo with two distinct cell types, the trophectoderm cells (T) and Inner Cell Mass (ICM). The cavity is marked with a C.
(Note the trophectoderm cells run the entire circumference of the embryo)

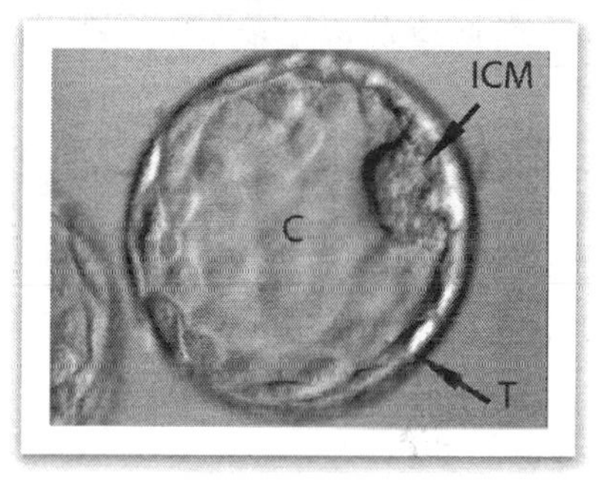

The overall structure of the blastocyst is very important, as is the presence of the two different cell types. Because the appearance of the embryo at this stage is dramatically different to earlier stages of development, blastocysts have their own separate grading system.

Blastocyst Grading

Blastocyst grading is more complicated because both cell types in the embryo are graded. What follows is just one example of a blastocyst grading system; your clinic may use a different system.

The blastocyst is designated a numerical value between 1 and 6. This reflects the degree of expansion or how much fluid the embryo has taken into the cavity. An embryo with a number one has just started to take in fluid and show signs of expansion and is still in the early stages. As an embryo takes in more fluid the grading system reflects this with a 2, 3 or 4 grade depending on how much fluid is inside the cavity and how big the embryo has grown. A number 5 is used to describe an embryo that is hatching out of its shell and a number 6 describes an embryo that is fully hatched out of its shell.

Next comes two letters, each one A B or C. The first letter describes the quality of the inner cell mass (or baby making) cells. If there are many tightly packed cells, the embryo will get an A grade. If there are less cells more loosely packed it will get a B grade and if there are few cells a C grade is given.

The grading system for the trophectoderm cells follows a similar pattern with A being the best and B and C being poorer quality meaning the cells are less organized.

A table of the blastocyst grading system can be seen below.

BLASTOCYST GRADING SYSTEM

Size or degree of expansion	What does the number mean?	Grade of trophectoderm	Grade of Inner Cell Mass
1	Slightly expanded	A = Good B = Average C = Poor	A = Good B = Average C = Poor
2	Cavity is more than half of the volume of the embryo	A, B or C	A, B or C
3	Full blastocyst with the cavity filling the embryo	A, B or C	A, B or C
4	Expanded blastocyst; the embryo has increased in size and the shell is thinning	A, B or C	A, B or C
5	Embryo is hatching out of its shell (partially hatched)	A, B or C	A, B or C
6	Embryo has fully hatched out of its shell	A, B or C	A, B or C

For example, an embryo with a grade of 4AA is fully expanded with good quality and well organized inner cell mass and trophectoderm cells. A grade of 3AB reflects a full blastocyst with a good inner cell mass and less well-organized or slightly poorer quality trophectoderm cells.

Examples of blastocysts with their grades can be seen below:

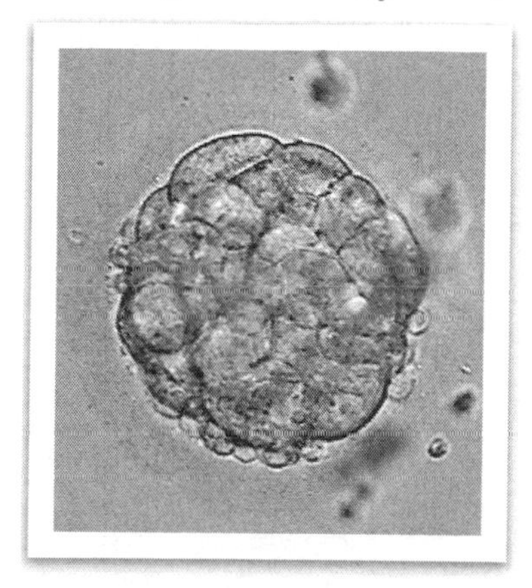

A blastocyst on day 5 of development
Early blastocyst, not yet able to grade

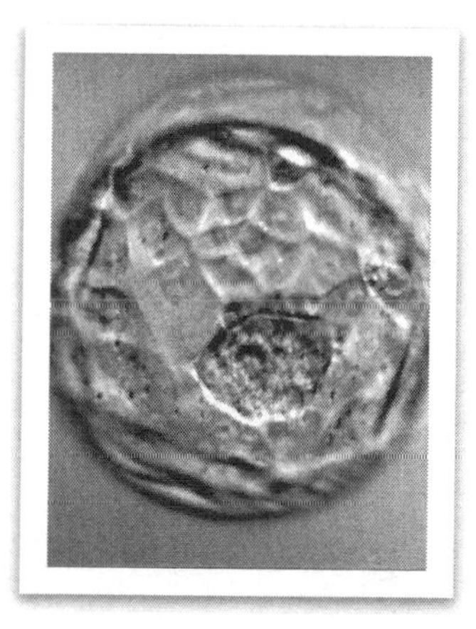

A blastocyst on day 5 of development
Grade 4AA

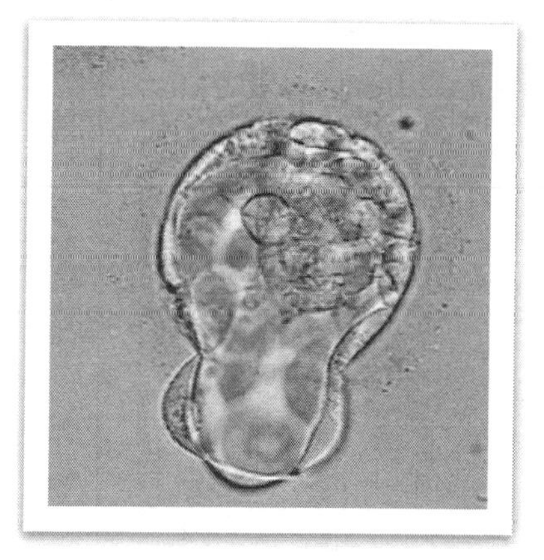

A blastocyst on day 5 of development
Grade 5AA, hatching

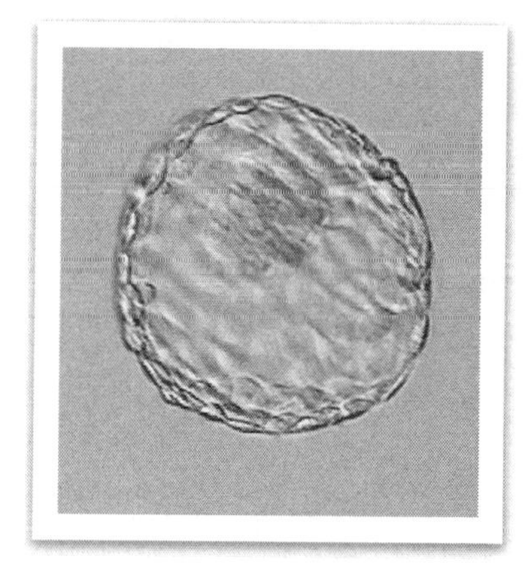

A blastocyst on day 6 of development, fully hatched
Grade 6AA

Hatching

From the beginning, each egg has a shell around it called the ZONA PELLUCIDA. On the photographs this appears as a shadow or halo around the eggs and embryos. The shell has two important functions; firstly it allows only one sperm to enter the egg during fertilization thereby maintaining the genetic integrity of the embryo. Secondly it holds together all the cells at the early cleavage stages of development.

By the blastocyst stage, the embryo is getting ready to implant in the uterus. This initially involves the expression or "showing" of molecules on the cell surface of the embryo, which recognize and bind to molecules on the cell surface of the uterine cells.

Normal blastocysts implant on day 6 or 7 of development by hatching out of the shell and coming into contact with the cells of the uterine lining.

In order for these molecules to meet and bind to the uterine lining, the embryo must first hatch out of its shell. Usually this happens on late day 5 or day 6 of development when the embryo takes fluid into the cavity and become too big for the shell to take the strain of the growing embryo. The blastocyst pulses, contracting and expanding until it gradually squeezes out of a hole in the shell. The previous photographs show embryos that are hatching and hatched.

Assisted Hatching

If the shells of the embryos are thicker than usual it may be difficult for blastocysts to hatch naturally. If this is the case, the embryonic cells would not be able to come into contact with uterine lining and implantation would fail.

Several studies have shown the benefit of ASSISTED HATCHING by which a hole is made mechanically or chemically in the shell of the embryo prior to embryo transfer. Once the embryos are hatched in the lab there is no chance of them being trapped inside their shells after they are transferred.

Assisted hatching can be done in several ways:

- Laser
- Partial Zona Dissection (PZD) or cutting the shell with micro-tools
- Acid Tyrodes which is a chemical that dissolves away part of the shell

Assisted hatching is not necessary for everyone and some studies have shown no benefit from hatching embryos when comparing two similar groups of patients. It is, however, a standard practice at most clinics and may be used for patients who:

- Have a raised FSH level
- Are over 38 years old
- Have poor quality embryos
- Have thickened shells around the embryos
- Have previously failed attempts at IVF
- All frozen/thawed embryos

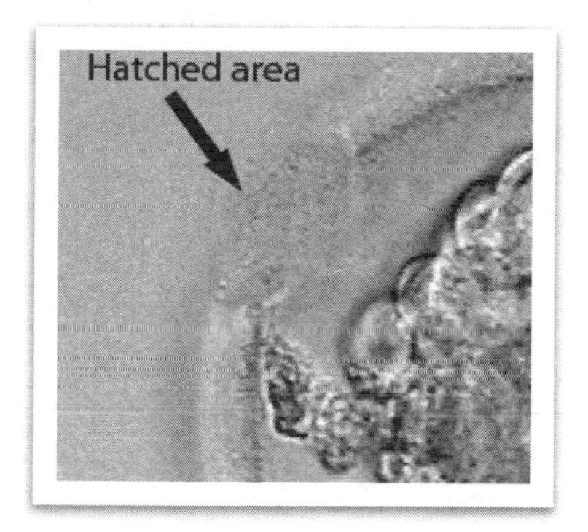

The embryos are now ready for transfer into the mother's uterus.

Embryo Transfer

Embryo replacement or transfer is performed by loading the embryos into a very fine catheter (or tube) and inserting it through the cervix to the uterus. The embryos are gently expelled from the catheter with a syringe and placed at the top of the uterine cavity. This can be done with or without ultrasound.

Embryo transfer should be painless and anesthesia is not necessary for the majority of people, however, you may be offered a muscle relaxant such as Valium to minimize uterine contractions following the procedure.

A common question we are asked is "Will my embryos fall out?" The answer is "No. They will not fall out!" Textbook illustrations that show uterine cavities as big open spaces are completely inaccurate. In reality the two walls of the cavity are pressed tightly together and there are lots of microscopic folds in the surface. This, coupled with sticky secretions inside the uterus, makes a safe place for your embryos to stay firmly put!

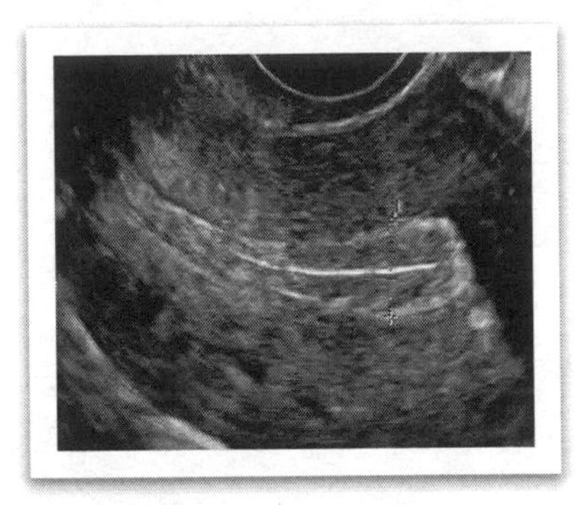

Ultrasound of an embryo transfer, the white line shows the path of the catheter.

WHEN TO TRANSFER: PROS AND CONS

Embryo transfer can be carried out on any day of development, although most clinics do so on either day three or day five of development. The primary benefit of growing the embryos in the lab as long as possible is that with each passing day there is a bigger variation in the quality of the embryos. This allows the best ones to be more easily identified.

Imagine a race with 20 runners. At the starting line they are all at exactly the same place; no winners and no losers. This is equivalent to day 1, when we have fertilized eggs and they all look the same. The runners in the race stay quite close together at the beginning but the longer the race goes, the stronger runners stream ahead of the slower ones and eventually there is a winner. Running the whole race would be equivalent to leaving the embryos to day 5 and choosing the leading blastocysts for transfer. You can see there is a benefit in leaving them in the lab until day 5 in order to choose the best ones when we have more than enough to choose from.

Now what would happen if there were only 3 runners in the race and you had to choose the top three? This would be easy to do earlier than the finish line. In other words, if there are only 3 embryos in total and you want to transfer 3 embryos, the same ones will be transferred on any given day; you do not benefit by leaving them to day 5 in the lab.

If we have just the right amount of embryos from the start or some clearly stop growing early on, it is possible to choose the "right" ones on day two or day three. In fact some people suggest that they might grow better inside the body than they would if they had been kept in the laboratory incubator.

There are, however, pros and cons to doing embryo transfers on either day.

By leaving the embryos to day five, the embryologist is better able to choose the best ones, giving a higher implantation rate while transferring fewer embryos. This gives the best chance of a pregnancy and reduces the chances of a high order multiple pregnancy (triplets or more, which are very high risk pregnancies).

The downside to doing blastocyst transfer is that some people will not have an embryo transfer at all. The embryos may all stop growing some time before day five. This of course is a devastating situation to face when you have made it so far through the IVF cycle. The only positive aspect of this is that you don't have to wait two weeks until the pregnancy test, doing bed rest and painful injections to find out that they didn't grow.

DAY	PROS	CONS
Day 3 ET	♦ Good chance of having a transfer ♦ Embryos may grow better inside the body ♦ Freezing on day three means more embryos frozen	♦ Transfer more embryos ♦ Greater chance of a multiple pregnancy ♦ Embryos would be in the fallopian tube at this stage, not the uterus ♦ Poorer quality embryos may be frozen ♦ Lower quality or "wrong" embryos may be transferred
Day 5 ET	♦ Transfer the best embryos, increased pregnancy rates ♦ Transfer less embryos ♦ Physiological conditions- embryos would be in the uterus at this stage ♦ Reduce the chances of multiple pregnancy ♦ If the embryos don't grow, you don't have the two week wait to find out, less medication and restrictions ♦ Good quality blastocysts frozen	♦ The embryos may not grow to blastocyst, (usually due to abnormal genetics) ♦ There is a chance of **no** transfer ♦ Less embryos frozen ♦ May lose some embryos that would have grown inside the body

In an attempt to reduce the number of multiple pregnancies from IVF, some clinics are turning exclusively to blastocyst (day five) transfer. The lab has to have an exceptionally good culture system that they trust to grow embryos to the blastocyst stage as well as they would grow inside the body. If this is the case, it allows the embryo selection to be made at the end of the "race" and increases the pregnancy rates for those who have an embryo transfer.

WHICH EMBRYOS TO TRANSFER?

The easy answer to this question is that we choose the best ones!

On any given day of culture we expect the embryos to have a specific appearance. On day two and three they should be 2, 4 or 8 cells, a morula on day 4 and at the blastocyst stage by day 5 or 6. When we do the embryo transfer it is important to choose only embryos that look viable given the day of culture. For example, we don't want to transfer an embryo on day 5 that looks like a day-3 embryo, which has surely stopped growing and won't implant and grow.

Take a look at the following embryos on day three of development:

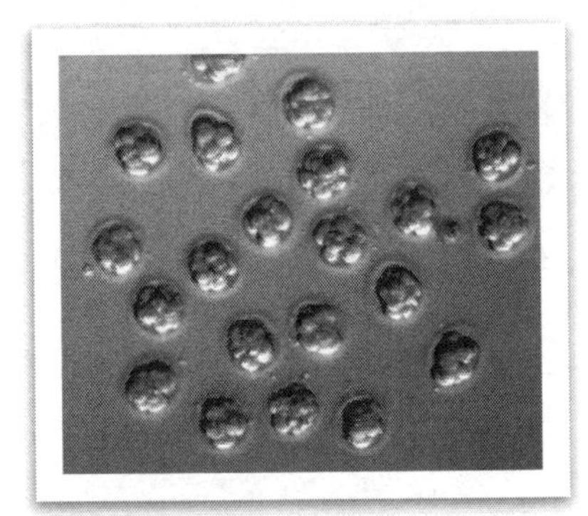

Here are the same embryos on day five of development:

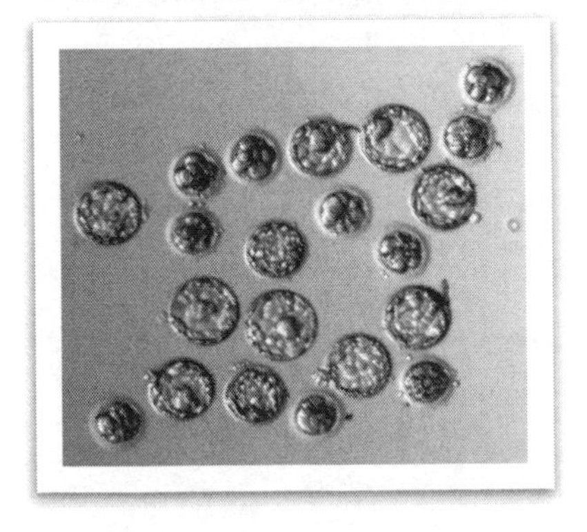

You can see they look very similar on day 3 and more diverse on day 5. By waiting until day 5 the ones with a better chance of success are chosen.

Now look at these embryos on day three, their overall quality is poor.

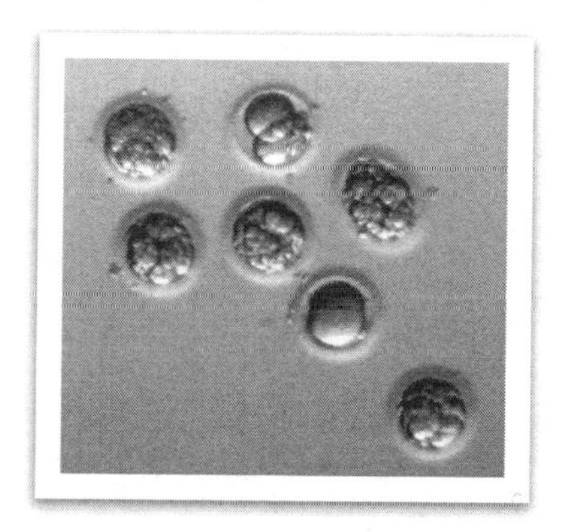

In this case there are two options.

1. Transfer on day three without knowing if they are going to make it to blastocyst.
2. Wait until day five and see if they make it to blastocyst risking the chance of not having a transfer.

If the lab has a good culture system, the outcome will probably be the same if you transfer on either day; they will hopefully continue to develop but they may not in either case. The difference is the disappointment of not having a transfer versus knowing sooner and stopping the medication and restrictions. It's not always an easy decision to make and you should follow the guidance of the doctor and embryologist.

In the above case, we waited until day five and transferred an expanded blastocyst and an early blastocyst. The patient is currently pregnant.

A WORD ABOUT THE EMBRYOS CHOSEN FOR TRANSFER

Because it takes two days for the embryos to change from the 8-cell stage to the blastocyst stage and is quite a slow process, there is a chance that the embryos chosen for transfer could be at an in-between stage. If we look at the embryos on day 5 and see only morula stage we usually go ahead and transfer the best of these. Although this isn't ideal for day 5, the most important thing is that the embryos have changed and developed since day 3 and still look viable. Pregnancies occur quite often following the transfer of morula stage embryos and in fact the body may give them the extra boost they need to get to the next stage of development.

HOW MANY TO TRANSFER?

Due to age related changes in fertility, the number of embryos transferred is directly related to the age of the mother or egg donor. We try to strike a balance between achieving a pregnancy and reducing the chances of triplets or more.

When deciding how many to transfer, your doctor will take into account the following:

1. Age of the female partner or egg donor
2. Stage at which the embryos are transferred (more embryos are transferred on day 3 compared with day 5)
3. Quality of the embryos
4. Number available
5. History and diagnosis
6. Previous treatment outcomes

The American Society for Reproductive Medicine (ASRM) has set *guidelines* for the number of embryos to transfer depending on age and situation. These are not laws and are used to help counsel patients. More than the recommended number of embryos can be transferred depending on the specific circumstances of each case.

The guidelines from the ASRM can be seen below:

GUIDELINES FROM THE ASRM

Age	Number of Cleavage stage embryos (Day 2 or 3)	Number of blastocysts (Day 5 or 6)
<35 Favorable conditions*	1–2	1
<35 Less favorable conditions**	2	2
35–37 Favorable conditions	2	2
35–37 Less favorable conditions	3	2
38–40 Favorable conditions	3	2
38–40 Less favorable conditions	4	3
41–42 Favorable conditions	5	3
41–42 Less favorable conditions	5	3

* Favorable conditions are first IVF cycle, good embryo quality, excess embryos available for freezing or previously successful IVF cycle.
** All other situations not listed above. There are no guidelines for women over the age of 41. The number of embryos transferred should be under the guidance of the individuals' doctor.

NOW WHAT?

After the embryo transfer you will be given instructions before leaving the clinic. These instructions will include what medication to take, when your pregnancy test is going to be and other lifestyle restrictions such as bed rest, no alcohol, caffeine or exercise. Clinics vary in the extent of restrictions following transfer and as a rule of thumb it is best to take it easy at least until the pregnancy test.

The pregnancy test will be approximately two weeks after your egg collection. On that day you will most likely have a blood test to detect the level of hCG (alternatively a urine test may be performed). HCG is a hormone that is produced by the pregnancy, so a positive level will indicate that the embryo(s) have implanted. A follow up test two days later checks if the pregnancy is progressing normally; the hCG level should approximately double every 48 hours.

Unfortunately, not all embryos develop normally and some pregnancies can end in miscarriage. Your Reproductive Endocrinologist will follow the hCG blood levels early in the pregnancy followed by one or two ultrasounds between 6 and 10 weeks of pregnancy.

Whatever the outcome of your fertility treatment, it is very important to stay on all prescribed medication and follow your doctor's instructions. A normal pregnancy will be followed until approximately 10 weeks gestation at the IVF clinic, at which time you will be referred to a regular obstetrician for pre-natal care.

An option for couples undergoing IVF is to have all the embryos genetically tested prior to embryo transfer. This can be done to reduce miscarriage rates by transferring only genetically normal embryos, to eradicate a genetic disorder from a family, or for gender selection of the offspring. In the next chapter we will look at how genetic testing is carried out and for whom it is most beneficial.

Possible Outcomes From an IVF Cycle

Outcome	Description
Pregnant	Normal levels of hCG doubling appropriately followed by ultrasound at 6–7 weeks detecting a heartbeat and a growing gestational sac.
Not pregnant	A negative or undetectable level of hCG. You will be told to stop all medication and given a follow up appointment to discuss the cycle and formulate a plan for the future.

Chemical pregnancy	An initial positive hCG level followed by an abnormal rise or a fall in hCG levels (pregnancy loss happens before a sac is detectable by ultrasound).
Blighted Ovum	An initial positive hCG level followed by a normal or abnormal rise in hCG levels. Ultrasound reveals an empty implantation sac and no heartbeat is detected. Follow up ultrasounds will be done to confirm that the pregnancy is non-viable and you will then stop all medication.
Ectopic	An ectopic pregnancy is when the embryo grows outside the uterus, usually in the fallopian tube but may also be on the cervix, ovary or bladder. This is a potentially dangerous situation and must be followed carefully by the doctor. The hCG level may start out low and double slowly but this is not always the case. Ectopic pregnancies can be detected by early ultrasound and are treated either with medication or surgery; (they are not viable pregnancies)
Miscarriage	The term miscarriage is used to describe a pregnancy loss. This can detected by ultrasound and can happen at various stages of a pregnancy. A dilatation and curettage (D&C) can be performed or the miscarriage may happen naturally. The most likely cause of a miscarriage is a genetic abnormality of the fetus. Other reasons include blood clotting or immune system disorders and hormonal imbalances.

5

Genetic Screening of Embryos

Pre-implantation genetic screening (PGS), also known as aneuploidy testing, is a term used to describe a genetic test carried out on cleavage stage (day 3) embryos prior to transfer. PGS does not diagnose any specific diseases but looks more generally at the genetic makeup of the embryo. This test counts the number of chromosomes present in a single cell taken from the embryo to ensure there are 46 chromosomes. No more and no less.

Pre-implantation genetic screening is not for everyone and has not been shown to improve pregnancy rates for the general population undergoing IVF. However for women with a history of recurrent miscarriage, PGS has been shown to reduce the miscarriage rate. PGS can also be beneficial for women of advanced maternal age and couples who have had repeated failed IVF cycles.

In order to understand genetic testing of embryos we must first grasp some basic cell biology. Each cell in the body has 46 chromosomes, except eggs and sperm which each have 23. The egg and sperm combine to give the offspring a total of 46 chromosomes—we inherit half of our chromosomes from our father and half from our mother. The chromosomes act like instruction books and control the function of the cells.

If an egg or a sperm is missing a chromosome then embryo created from that egg will be also missing a chromosome and a whole set of instructions. If there is an extra chromosome in an egg, the embryo will be abnormal by having twice as many copies of those instructions than it should; an example of this is Down's Syndrome.

It is possible to test each individual embryo created through IVF and count the number of chromosomes present. This is done by taking a single cell from the embryo on day 3 of development and running a test on that cell to look at its genetic make-up.

The removal of a single cell from the embryo at this stage does not affect its development in terms of making it abnormal. The embryo can easily compensate with a few cell divisions. In good hands, the biopsy of an embryo has little effect on its growth, although it may slow down slightly.

Embryo biopsy on day 3 of development

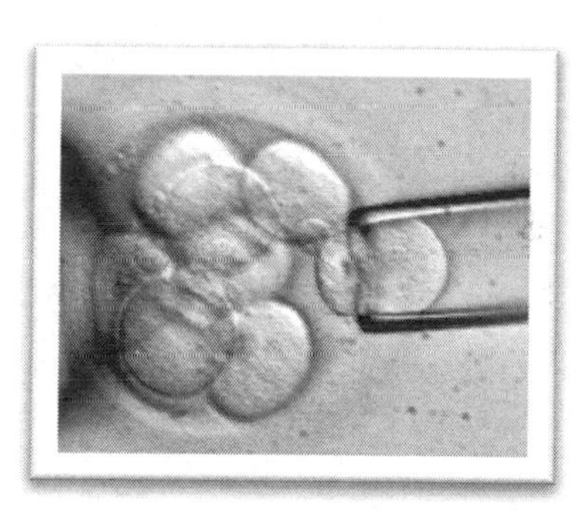

Following biopsy, fluorescent dyes are added to the cell and a different color is used for each chromosome. When viewed down the microscope the colors shine as bright spots. If there are two copies of any chromosome there will be two bright spots of that specific color, if there are three copies of that chromosome there will be 3 bright spots etc. This is called fluorescent in-situ hybridization or FISH.

One limitation of FISH is that if the chromosomes are very close together or overlapping there may appear to be only one, which would give a false report of an abnormal embryo, although this is quite rare.

PGS does have some limitations. The rationale behind the test is the presumption that each cell in the embryo is genetically identical. In other words, if you test one cell from the embryo you know what all the other cells are. There have been reports of embryos exhibiting mosaicism, where not all the cells are identical. This leaves open the possibility of misdiagnosis and the discarding of normal embryos or the transfer of abnormal embryos.

Because most genetically abnormal embryos stop growing before the blastocyst stage or fail to implant, most people do not have increased pregnancy rates following PGS. The outcome of the IVF cycle is likely to be the same with or without the testing. The main difference is that by doing PGS you have the information sooner rather than later. PGS can help women who are struggling to decide whether to continue to use their own eggs after several failed IVF cycles.

PGS Pros	PGS Cons
♦ Reduction in miscarriage rate ♦ Provides information to help you make decisions about future treatments ♦ Can help women of advanced maternal age or with repeated implantation failure identify genetically normal embryos	♦ No improvement in pregnancy rate for the general population ♦ Embryo may be damaged during biopsy ♦ Results may not be 100% reliable ♦ May be cost prohibitive ♦ Theoretically does not eliminate the need for prenatal testing ♦ Unable to test for all chromosomes (maximum of 12)

If you do decide to do PGS, the biopsied embryo cell will be sent for analysis and the embryos will stay in the incubator at the IVF clinic. Results from the testing take 1 to 2 days and are usually available by day 5 of embryo development in time for a fresh blastocyst transfer.

The embryos chosen for transfer have to be both genetically normal and have a normal pattern of growth. There is no point transferring a 4-celled embryo on day 5 which has clearly stopped growing even if the genetic analysis came back normal. In other words, there is always a chance of not having an embryo transfer regardless of the PGS results.

The following table is an example of a PGD report. You can see in some cases there are two copies of each chromosome and normal sex chromosomes (XX is a female, XY is a male). These are the normal embryos that would be chosen for transfer, depending on the grade of the embryo and whether it has grown to blastocyst.

AN EXAMPLE OF A PGD REPORT FOR ANEUPLOIDY

Emb.	XY	8	13	14	15	16	17	18	20	21	22	Interpretation
1	XX	2	2	2	2	2	2	2	2	2	2	NORMAL
2	XXY	3	3	3	3	3	3	3	3	3	3	POLYPOID
3	-	-	-	-	-	-	-	-	-	-	-	No Nucleus
4	XX	3	2	2	2	2	1	1	1	2	2	Complex abnormal
5	XX	2	2	2	2	2	2	2	2	2	2	NORMAL
6	X	2	2	1	1	1	3	1	2	3	2	Complex abnormal
7	XY	2	2	2	2	2	2	2	2	2	2	NORMAL
8	XX	2	2	2	2	2	2	3	2	2	2	Trisomy 18
9	XX	2	2	2	2	2	2	2	2	1	2	Monosomy 21
10	XY	2	2	2	2	1	2	2	2	2	1	Monosomy 16, 22
11	XX	2	2	2	2	2	2	2	2	2	2	NORMAL
12	XY	2	2	2	2	2	2	2	2	3	2	Trisomy 21

PGS can also be used to detect translocations, which are changes in the structure of the chromosomes. It is quite rare, but does account for some cases of recurrent miscarriage or implantation failure. It is estimated that 1:700 people carry a translocation, so if you have suffered multiple miscarriages or multiple failed IVF cycles it might be a good idea to talk to your doctor about checking the structure of the chromosomes for translocations in the female and the male partners. This testing is called karyotyping.

It is possible to have a sperm sample tested for 5 chromosomes using FISH. Reprogenetics, LLC offers this service and counts the chromosomes in 500 sperm. Although normal sperm cannot be selected for fertilization using this technique, it does provide patients with information about the overall contribution of the sperm to the genetic makeup of the embryo and may help with lifestyle changes and future treatment decisions.

PRE-IMPLANTATION GENETIC DIAGNOSIS (PGD)

While PGS counts the number of chromosomes, PGD is used to identify a *specific* disease in a pre-implantation embryo.

Genetic diseases are often caused when a gene mutates or the instructions have become misspelled on part of the DNA. This causes the cell to malfunction and not produce the correct proteins needed to do a specific task in the body. Genetic diseases of this type are called SINGLE GENE DISORDERS.

A person who carries this type of genetic disorder can undergo a cycle of IVF. This presents an opportunity to eliminate this disease from your family and future generations. Accuracy of the results is close to 100% and although PGD does carry some risk of damaging the embryo during biopsy, this is offset by the peace of mind in knowing you are carrying a baby without a specific disease.

Here are several of the diseases PGD can be used to detect:

- Cystic fibrosis
- Sickle cell anemia
- Tay-Sachs disease
- Myotonic dystrophy
- Duchenne and Becker muscular dystrophies
- Fragile X syndrome
- Spinal muscular atrophy
- Huntington's
- Hemophilia

For PGD testing, a single cell is removed from each embryo on day 3 of development and the DNA is multiplied many times. A specially made probe is then used on the specific region of DNA (or gene) that carries the disease mutation. The results confirm whether a particular embryo is carrying the defective gene or has normal genetics.

COMPARATIVE GENOMIC HYBRIDIZATION (CGH)

CGH has been heralded as a major infertility breakthrough. It is a genetic test of eggs and embryos that offers more information and possibly more precise information than standard genetic testing has in the past.

During CHG testing the DNA from the embryo is labeled or "tagged" with a green fluorescent dye. The DNA of a cell known to be completely normal is labeled with a red fluorescent dye. The two samples are added together and the DNA from both combines and becomes a neutral color. If either green or red is still present after the samples are added together this means that there is either an extra chromosome or a missing chromosome in the test sample. Depending on the color and the amount of fluorescent dye present, it is possible to determine which chromosome is aberrant.

The advantage of doing CGH testing over traditional FISH testing is that all 23 pairs of chromosomes can be tested (versus only 12 chromosomes using FISH) and the results are more accurate when testing for aneuploidy.

A downside to CHG is that the test cannot detect polyploidy (when the embryo has several sets of chromosomes). This is because the test looks for deviations between the chromosomes. Since all the chromosomes are duplicated in polyploidy this situation would be interpreted as normal.

Also, due to technical restrictions at this time, CGH is less accurate in determining gender than FISH, at 85% accuracy versus 100% for FISH.

Embryo testing can be performed on either day three or day five. The advantage of doing blastocyst biopsy is that several cells can be tested thereby reducing the risk of a false result due to moisaicism. However, if embryos are tested on day 5, they have to be frozen because the results aren't available in time for a fresh transfer. This may change with time and better technology.

CGH has many of the same negative aspects as traditional PGS/PGD testing. It is an invasive test that may damage the embryo during biopsy, there may be false readings and the results may not increase pregnancy rates for most people going through IVF.

The future of genetic testing is likely to include micro-array CGH which will provide information about chromosome imbalances, including some that may be too small to detect by traditional PGD. This test may be able to detect behavioral disorders such as autism.

Choosing the Gender of Your Baby

Pre-implantation genetic testing can be used to detect the gender of the embryos when a baby of a certain sex is desired. This is a controversial use of reproductive technology, however, most clinics do offer this as a service. Some clinics may have restrictions on which families they allow to choose gender. They may require a couple to already have children and want a child of the opposite sex; this is called family balancing.

During an IVF cycle, the embryos are biopsied and sent for FISH analysis in exactly the same way as aneuploidy testing. A certain number of chromosomes could also be tested at the same time to ensure genetic normalcy of the embryo as well as the gender.

At the time of the embryo transfer you will be told the results of the genetic analysis and the quality of the embryos. It is possible that there may not be good quality embryos of the desired gender and so you would have to decide whether to transfer the alternate gender or lower quality embryos in that case. The downside of doing PGS is that there is a potential of disrupting the embryo growth by doing the biopsy; high cost and the chance of misdiagnosis are also considerations.

An alternative to PGS for sex selection is to put the sperm though a machine that separates out the X and the Y sperm, followed by an IUI or IVF cycle. This separation is based on the significant difference in size between the X and the Y chromosomes. Sperm carry an X **or** a Y chromosome, which determines the gender of the offspring.

Using flow cytometry the sperm are enriched or "sorted" according to the chromosomes they carry. Using this method there is a 93% chance of having a girl and an 82% chance of having a boy (based on the most recent data from Microsort).

A company called Microsort performs sperm sorting. There are two locations in the US, Fairfax, Virginia and Laguna Hills, California. For a good chance of conception with an IUI, it is best to travel to one of these locations and produce a fresh sample for sorting and insemination.

It is possible to freeze a sperm sample at your local IVF clinic and have it shipped to a Microsort facility where it will be thawed, sorted, refrozen and shipped back. Because of the multiple freeze thaw processes, this would only be recommended for good quality specimens and you will require an IVF cycle with ICSI to use this sperm.

According to the latest data from Microsort, the average cumulative IUI clinical pregnancy rate is 16.6% and the overall pregnancy rates with IVF/ICSI is 34.7%

The fees for Microsort are approximately $3000 for the sperm enrichment, not including consultation fees and charges for other treatments (such as IUI or IVF/ICSI).

NATURAL METHODS FOR GENDER SELECTION

Natural methods of gender selection include timing of intercourse according to ovulation, and changes in diet and supplements (vitamins and minerals). The male sperm are said to travel faster and burn out sooner whereas the female sperm are slower and last longer. For a girl you should have intercourse a few days before ovulation and for a boy have intercourse right around the time of ovulation. This method is not scientifically proven.

Some scientific evidence supports dietary changes having an influence on the gender of the offspring, but the differences are very slight and only deviate a few percent from the national average.

These methods include changing the acidity in the body and changing calorie and salt intake during the pre-conception months. For a girl, the woman should lower the salt and reduce the calorie intake for at least 3 months prior to conception. For a boy, do the opposite, higher dietary salt and more calories. Historically, more boys are born in times of plenty.

An alternative is to keep your fingers crossed and love the one you get!

6

Cryopreservation: Fertility on Ice

Cryopreservation is the freezing of biological material for future use. At the fertility clinic, cryopreservation is routinely used to store sperm and embryos, and more recently, human eggs.

Eggs, embryos and sperm frozen in liquid nitrogen can be stored indefinitely; there is no biological activity or degradation at such low temperatures. Babies have been born from embryos and sperm that have been stored for over 10 years. Evaluation in several studies of children born following cryopreservation has not found any increased risk of developmental delays or birth defects in children born from frozen-thawed embryos.

Embryo Freezing

BASIC PRINCIPLES

The primary advantage of freezing embryos is it allows the maximum potential for pregnancy from one IVF cycle; if there are excess embryos of good quality following the fresh embryo transfer these can be preserved for future use. In some clinics the pregnancy rates from frozen-thawed embryo transfer are almost as high as the rates achieved with fresh embryos. However, due to the high rate of embryo attrition during IVF, less than half of the cycles will have sufficient embryos to freeze following the fresh transfer. If you do not have spare embryos, try not to be too disheartened and remember the best chance of success is with the fresh embryos.

If a woman is at risk of developing severe Ovarian Hyper-Stimulation Syndrome (OHSS), which is exacerbated by pregnancy, then all the embryos from an IVF cycle may be frozen. In this situation there is no transfer and all the embryos are cryopreserved for a period of at least two to three months before the embryos are thawed and transferred. This gives an opportunity for the woman to fully recover from OHSS and have the embryos placed into a healthy uterus.

Unexpected conditions can arise around the time of an IVF cycle such as illness, poor endometrial development, breakthrough bleeding or a premature rise in progesterone. Cryopreservation can be used to postpone the transfer until a later date when the conditions are more favorable. This allows flexibility and the knowledge that the IVF cycle is not lost if a fresh transfer doesn't take place.

STAGES OF EMBRYO DEVELOPMENT AND CRYOPRESERVATION

Embryos can be frozen at various stages of development. These are most commonly:

- Day 1 (2PN)
- Day 2 or 3 (Cleavage stage)
- Day 5 (blastocyst stage)
- Day 6 (blastocyst stage)

Each clinic has its own freezing preferences, which are usually based on the stage at which embryos are typically transferred.

There are pros and cons to freezing at the different stages of development. When embryos are frozen early in their development (for example at the 2PN or cleavage stage) there will be *more* embryos frozen but they are of undetermined quality. When embryos are frozen on day 5 or 6 of development, fewer embryos are frozen but they have already proven themselves to be the best quality and able to form blastocyst stage embryos. In other words, quantity does not always equal quality, as with the fresh IVF cycle embryos.

When embryos are frozen early in their development, it may be necessary to thaw most, or all, of the embryos to get enough of good quality for transfer. With blastocyst freezing, it is common to thaw only the number of embryos desired for transfer.

When we freeze *all* the embryos, for example if someone has OHSS or for fertility preservation, we can freeze at any stage of development. While you want to have only viable embryos in the freezer, it can be quite disheartening for someone who has created a high number of embryos, to have only a small number cryopreserved. This is especially true if there was no fresh embryo transfer. Whatever stage the embryos were frozen, the outcome of the frozen embryo transfer cycle will likely be the same.

Remember, most clinics only freeze good quality embryos; less than half of all IVF cycles have spare embryos for freezing due to the attrition rate of the embryos created. Don't be too disappointed if you don't have frozen embryos, the best chance of pregnancy is with a fresh transfer.

METHODS OF EMBRYO CRYOPRESERVATION

There are currently two methods of embryo cryopreservation, **slow freezing** and **vitrification**.

Both methods involve placing the embryos in a cryoprotectant fluid that protects them during the freezing process. The degree of success of freezing and thawing embryos comes from the efficient removal of water from the cells. When water freezes it expands and forms ice crystals that are destructive to the delicate cellular components. Cryoprotectants have been specially designed to get inside the cells and replace the water. The cryoprotectant does not form crystals and can safely withstand the drastic reduction in temperature. Not all cells can be successfully dehydrated and some embryos are more prone to damage than others. Also, some embryos will survive the thaw while others will not. It is unpredictable which embryos are more robust and the results can only be seen on the day of the thaw.

Occasionally, some cells within a single embryo are alive and some are not. This can make it difficult to grade frozen/thawed embryos accurately. In general, if an embryo has at least 50% of its original cells it is worthwhile to transfer it.

During the slow freezing process, the embryos are first dehydrated through a series of alcohols and then placed in a special machine that reduces the temperature by one degree per minute or less. It can take over an hour to reach -80 degrees centigrade at which time the embryos (in their storage straws) are plunged into liquid nitrogen.

Vitrification is a more modern freezing technique. The embryos are placed in strong solutions of dehydrating media and then plunged directly into liquid nitrogen in their storage vessel. This method takes just a few minutes to complete.

Many studies have shown benefits of vitrification over slow freezing. Although both methods have produced healthy pregnancies, it is the general consensus among the medical community that vitrification produces superior survival rates and subsequently higher pregnancy rates than slow freezing embryos.

Embryos historically, have been frozen in batches of two or three; however, with the high survival rates of vitrified embryos it is quite common for them to be frozen singly. This way only the exact number of embryos desired for transfer need to be thawed.

THAWING OF EMBRYOS

For thawing, embryos are slowly warmed to room temperature and washed several times to remove the cryoprotectant solutions. It is the reverse of the freezing process. They are then kept in the incubator for observation for at least an hour before embryo transfer to ensure viability. Approximately 70–80% of embryos are expected to survive the thaw.

If the embryos were frozen at the blastocyst or cleavage stage of development they can be transferred into the uterus on the same day as the thaw. If they were frozen at the 2PN stage they will be cultured in the lab for up to five days to optimize selection of the best ones for transfer. There is always the possibility that none of the embryos will survive the thaw in which case the transfer will be cancelled.

Re-freezing thawed blastocysts is uncommon and usually unsuccessful. However, if embryos are frozen early in their development and cultured to the blastocysts stage, any extra, good quality blastocysts may be re-frozen.

Most clinics perform assisted hatching on frozen/thawed embryos as the freezing process can harden the shell around the embryo making it more difficult for the embryo to hatch naturally.

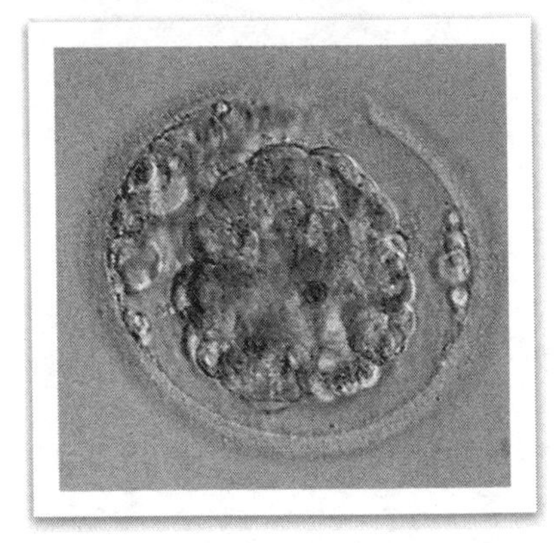
A previously vitrified day 6 blastocyst, immediately following thaw

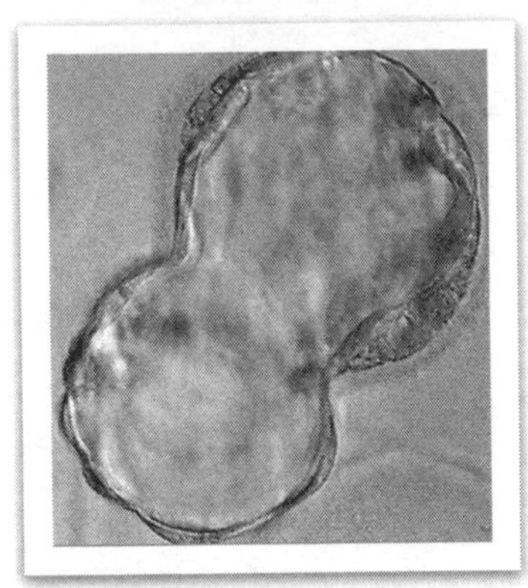
The same embryo 3 hours later, the blastocyst has fully re-expanded and is hatching out of its shell.

FROZEN EMBRYO REPLACEMENT CYCLE

There are several ways the frozen/thawed embryos can be replaced, these include:

- During a hormone replacement cycle
- During a natural cycle, without drugs
- During a stimulated cycle with Clomid or FSH injections

HORMONE REPLACEMENT CYCLE

Most frozen embryo transfer cycles are performed using hormones to prepare the uterine lining. This usually begins with Lupron to suppress the pituitary gland and prevent ovulation from occurring naturally. This also ensures the uterus is in synchrony with the embryos.

Following two weeks of Lupron injections, estrogen is introduced to mimic the changes that normally take place during the menstrual cycle. A steady supply of estrogen for the first half of the cycle prepares the uterine lining by thickening and maturing the endometrium. During this phase, the woman will usually be monitored to determine the thickness of the uterine lining. Blood tests may also be carried out to check the level of estrogen.

Sometimes estrogen therapy may be prolonged to ensure maximum endometrial thickness or if the scheduling of the transfer needs to be changed. The length of estrogen supplementation is flexible and allows the timing of the embryo transfer to be changed; this can be less stressful for the patient compared with a fresh IVF cycle.

Once the uterine lining is sufficiently thick, progesterone supplementation begins and Lupron is discontinued. Progesterone matures the lining and makes the endometrium receptive to embryo implantation. The stage of the uterus must match with the stage of embryo development in order for successful implantation. There is a limited window of time before the uterine cells change and are no longer able to support a pregnancy.

The timing of the embryo transfer will be according to the stage at which the embryos were frozen. For example, if the embryos were frozen at the blastocyst stage they will be transferred after 5 days of progesterone. If they are going to be transferred at the 8-cell stage, they will be transferred after 3 days of progesterone and so on. Progesterone supplementation will continue until the pregnancy test and beyond if the pregnancy test is positive.

STIMULATED CYCLE

In a stimulated cycle, the embryos are again transferred during the receptive phase of the cycle, however, the uterus has been prepared by taking Clomid or FSH to produce one or two follicles prior to transfer. HCG is given to induce ovulation and the precise timing of the embryo transfer is determined according to the synchrony of embryo development and uterine receptivity.

NATURAL CYCLE

Frozen/thawed embryos can also be transferred during a woman's own natural menstrual cycle. This is usually advised only for women who have regular, predictable periods. The pregnancy rates are similar for natural cycle and hormone replacement cycles, however, the natural cycle frozen embryo transfers are more difficult to co-ordinate. In order to ensure the embryos are replaced during the "window of receptivity," the day of ovulation must be precisely determined. This can involve home ovulation predictor kits or measuring the levels of hormones in the blood. Sometimes supplemental progesterone is given during the second half of the menstrual cycle to support the endometrial lining during a natural cycle frozen embryo transfer.

If you did not get pregnant during the fresh IVF cycle and have good quality frozen embryos, a frozen embryo cycle may allow for a transfer into a more physiologically natural environment. This is especially true for people who had exceedingly high levels of estrogen and progesterone during the stimulated cycle or those who suffered from OHSS.

If the thawed embryos are good quality, pregnancy rates may be higher during natural cycles or during cycles where the hormone levels do not exceed that which occurs naturally.

PREGNANCY RATES FROM FROZEN EMBRYO REPLACEMENT CYCLES

Because not all thawed embryos will survive, it may be necessary to thaw several embryos to get two or three good quality ones to transfer, especially if the embryos were frozen early in their development.

The success rates of frozen embryo cycles depend on many factors; mainly the woman's age at the time of freezing and the number of embryos transferred. The number available, stage of development and the embryo quality also have an influence on the chance of success.

Overall, the chance of success with a frozen embryo transfer is lower than with fresh embryos. For this reason some clinics recommend replacing one more embryo than you did for the fresh transfer, especially if the fresh cycle was not successful. If the fresh IVF cycle was a success and the frozen embryo transfer is an attempt at a subsequent pregnancy, then you may want to be more conservative.

ETHICAL ISSUES

If you have frozen embryos in storage there may be ethical issues to consider.

- Who has ownership of the embryos if the couple divorces?
- What is the fate of the stored embryos if one or both parents die?

Some couples leave a provision in their will defining what should be done with the frozen embryos upon their death. Clinics will help you with these issues before treatment begins and will have the designation written as part of the cryopreservation consent form. To avoid confusion, it is a good idea to talk with your partner about these issues prior to treatment.

There are several options available for what to do with your remaining embryos if you still have embryos frozen once your family is complete. Your clinic may be able to guide you towards the right decision for your family.

Here are some options for your frozen embryos if you no longer desire more children:

- Allow the embryology staff to dispose of the embryos at the IVF clinic
- Take them home and have a private ceremony or prayer
- Have them transferred into the uterus at a time when pregnancy is highly unlikely
- Donate them for stem cell or other biological research
- Donate them to another couple for fertility purposes
- Continue long term storage in the event a Human Leukocyte Antigen (HLA) match is required for an existing child

Sperm Freezing

Due to the difference in cell size, sperm can tolerate freezing more readily than embryos and usually survive the thaw process very well. Samples can be frozen for men who are not available to produce a specimen on the day of egg retrieval or as a back-up plan for men who are anxious about producing a sample.

Surgically removed sperm can also be cryopreserved ahead of time, which allows an IVF cycle to go ahead with the knowledge that there is sufficient sperm to proceed. Also, donor sperm can be frozen and stored, allowing an adequate time period for quarantine and communicable disease testing (usually a minimum of 6 months).

QUESTIONS TO ASK

- What methods of freezing do you use, slow freezing or vitrification?
- At what stage of embryo development do you freeze?

- How are embryos batched (singly or in two's or more)?
- What are the pregnancy rates for frozen embryo replacement cycles?
- How long can we store the embryos at your facility?
- How much is the annual storage fee?
- What can we do with "spare" embryos after our family is complete?

Case Study

Age	35
Diagnosis	Male factor infertility
Process	Slow frozen
2006	Fresh IVF cycle, 2 blastocysts transferred, 6 frozen: early miscarriage
2007	Frozen embryo cycle, 3 thawed, 2 transferred, healthy baby born
2010	Frozen embryo cycle (shown below), remaining 3 thawed, 3 transferred, twins

Immediately following thaw, Blastocysts are collapsed due to the dehydration process.

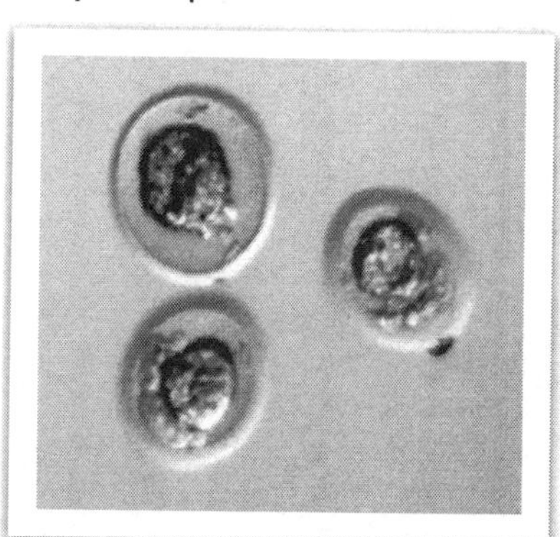

One hour after thaw.

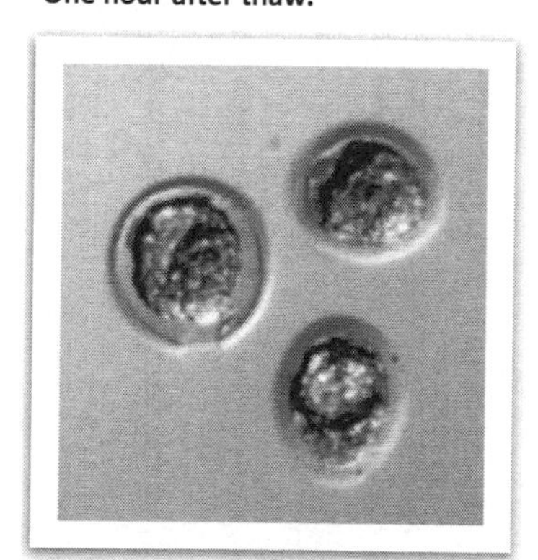

Assisted hatching

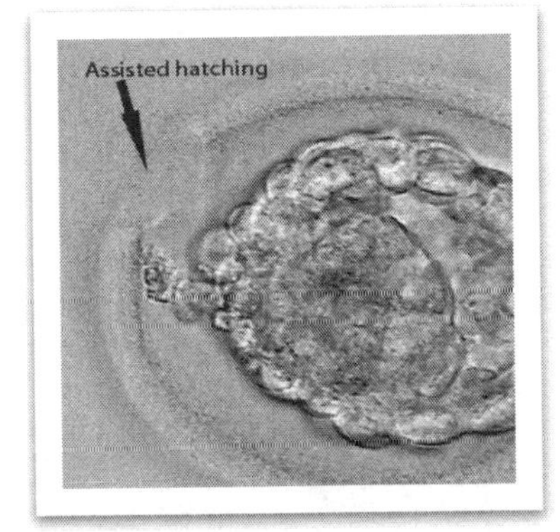

5 hours post thaw (at time of transfer)

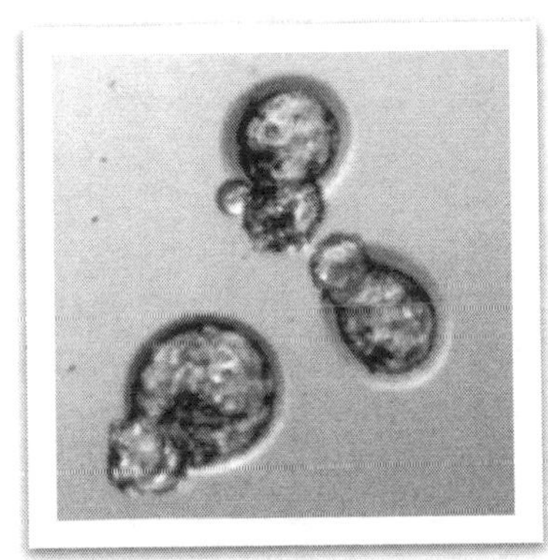

Following an early miscarriage with a fresh IVF cycle in 2006, this patient had a successful delivery in 2007 with a frozen-thawed embryo replacement cycle. Wishing to complete their family, all three remaining embryos were thawed and following their survival, all three were transferred. The rationale behind transferring three embryos was the couples' history, the fact that these were frozen embryos with slightly lower potential than fresh, and their willingness to accept a twin pregnancy. Initial hCG testing was positive followed by an ultrasound which revealed a twin gestation.

Fertility Preservation: Freezing Eggs

Egg freezing is a relatively new technology; it has previously been difficult to preserve eggs due to their delicate cellular make up. A new technology called vitrification has allowed advances in egg cryopreservation, although some clinics have had success using the slow freezing method.

A reason for freezing eggs is for fertility preservation, when a woman is not ready to have children but feels her biological clock ticking, or if a woman has a serious illness and treatment will leave her infertile. This is an option for some women with cancer who are about to undergo chemotherapy, although there may not be enough time to do a fully stimulated cycle before chemotherapy treatment begins. Fertile Hope is an organization that helps women with cancer by funding part of the IVF cycle. They can be reached at Fertile Hope, (888) 994-HOPE, www.fertilehope.org.

Eggs can also be frozen in the event that no sperm is available on the day of egg retrieval or if a couple has moral objections to freezing embryos.

EGG VITRIFICATION

Egg vitrification is similar to embryo vitrification and involves dehydrating the eggs in a series of cryoprotectants and then plunging them into liquid nitrogen in a very small volume. It literally means "turning to glass."

A stimulated cycle is usually required prior to egg freezing, (with the exception of some women who are about to undergo treatment for cancer which cannot be delayed). This involves four to six weeks of medication to stimulate the ovaries to produce many eggs at one time. These eggs are surgically removed once the follicles have reached full maturity.

The mature eggs are processed through solutions containing high concentrations of cryoprotectants. They are then quickly placed in a tiny drop of fluid that is barely larger than the egg itself, and placed onto a "leaf" that looks like a small spatula. The "leaf" is plunged into liquid nitrogen before being placed into a secure holder labeled with the patient's identifying information.

When embryo transfer is desired, the eggs are thawed and fertilized using ICSI before being cultured in the lab for several days and transferred into the uterus as embryos.

EGG FREEZING FAQ

How does vitrification of eggs work?	Freezing biological material can be challenging due to the formation of ice crystals. Water expands when it is frozen causing damage to the surrounding tissue if it is not fully removed prior to freezing. Any type of cryo-preservation involves dehydration of the tissue, and vitrification uses high concentrations of these dehydrating solutions. During the freezing process all the water inside the cell is replaced with the cryo-protectant solution, enabling the cells to be plunged into nitrogen without any ice crystal formation.

Is it safe?	To date, there have been approximately a thousand babies born from frozen eggs. These babies appear to be perfectly healthy and normal. When scientists begin using a new technique, we never know the long-term effects, however, historical use of egg/embryo freezing leaves us confident that this method is entirely safe.
Why is vitrification a better option for freezing?	Because vitrification uses higher concentrations of cryoprotectants, it is more efficient at removing the water from the egg cell. This, coupled with the faster freezing rate allows the eggs to be further protected from the damaging effects of the freezing process. The old way of freezing was slower and gentler but often did not fully dehydrate the cells and left the eggs and embryos susceptible to damage.
What is the success rate from thawed egg IVF?	The success rates from using frozen eggs depend on a variety of factors, most importantly the age of the woman at the time of freezing. Other factors include how many eggs are available and underlying fertility problems that may inhibit pregnancy. It is usually recommended that at least 10 eggs be stored in order to give the best chance of a viable pregnancy. In general, if the eggs survive the freeze/thaw process and fertilize normally they are just as likely to grow as embryos created from "fresh" eggs. The pregnancy rates will be slightly lower overall but egg freezing does have good prospects for people with normal fertility at the time of freezing.
How will it help success rates in donor egg programs?	It is possible that in the future, we will see a wave of egg banks that offer frozen eggs at a lower price than doing a fresh IVF cycle. Egg vitrification is not necessarily going to help success rates in the future as embryos from "fresh" eggs are more likely to implant than frozen; but, it may provide a more affordable and accessible option for some people. One positive aspect of using frozen eggs is that they are readily available and pre-screened so their use alleviates a lot of the waiting time and synchronization between the donor and recipient.

Can thawed eggs be refrozen as embryos?	Yes. Once we thaw the eggs and fertilize them we usually see a pattern of growth that is very similar to fresh eggs/embryos. When we get to day five of culture and transfer the best blastocysts, we freeze the spare embryos that are of good quality. The freeze/thaw success rate of these embryos is the same as embryos that came from eggs that have never been frozen.

7

Complications from Fertility Treatments

Complications from fertility treatments are quite rare, however, some people will experience side effects. It is advantageous to be informed and on the look-out for signs and symptoms to bring to your doctors attention. Complications can occur during the IVF cycle itself, during the pregnancy or with the offspring. Here are several of the most common risks involved with fertility treatments.

Ovarian Hyperstimulation Syndrome (OHSS)

Any patient undergoing ovarian stimulation is at risk of developing OHSS, which is one of the most serious side effects of fertility treatments. Mild OHSS occurs in around 30% of all treatment cycles and is considered quite a "normal" side effect of ovarian stimulation for IVF. The symptoms can be easily managed with over the counter analgesics, with dietary changes and increased fluid intake. Severe OHSS affects less than 1% of all women who undergo IVF treatment and can be managed well with medical help. Women at high risk of developing complications include those with polycystic ovaries or who have a high response to the stimulation drugs and produce many follicles. The symptoms of OHSS may be mild, moderate or severe and usually begin four to five days following egg collection. Mild or moderate OHSS usually resolves within a few days unless pregnancy occurs which may delay recovery. Because the symptoms of OHSS are exacerbated by pregnancy, sometimes a "freeze all" is the safest option. In this case, all the embryos will be frozen and a transfer delayed until full recovery two or three months later.

OHSS is caused by high levels of estrogen in the blood stream, which cause the ovarian blood vessels to become "leaky" to fluids. This means fluids leave the blood stream and collect in the abdominal cavity. Even though you are drinking plenty of fluids you can become severely dehydrated. As a result, the blood becomes thickened leaving you at risk for blood clots in the legs and lungs. The accumulation of fluid in the abdominal cavity can be very uncomfortable and cause bloating and sometimes breathlessness. In severe cases, the fluid can enter the lungs and cause respiratory distress. This is why it is very important to inform your doctor if you have any of these symptoms.

Mild OHSS	Moderate to Severe OHSS
♦ Mild abdominal pain ♦ Abdominal bloating and weight gain ♦ Mild nausea ♦ Vomiting ♦ Diarrhea ♦ Tenderness around the ovaries	♦ Rapid weight gain ♦ Severe abdominal pain ♦ Persistent nausea and vomiting ♦ Decreased urinary frequency ♦ Dark urine ♦ Shortness of breath ♦ Tight and enlarged abdomen ♦ Dizziness

If you have any of the symptoms of hyperstimulation, it is very important that you inform your doctor. He or she will monitor your condition and assess if any intervention is necessary. Treatment usually involves keeping you comfortable until the symptoms subside, which can take a week or two or longer if you are pregnant. For more severe cases the doctor may drain excess fluid or administer intravenous fluids.

WHAT YOU CAN DO:

- Take over-the-counter pain relievers, such as Tylenol.
- Reduce activities, no heavy lifting, and straining or strenuous exercise.
- Maintain light activity. Total bed rest may increase the risk of some complications.

- Elevate your feet when resting. This helps your body get rid of the extra fluid.
- Avoid sex until you feel better.
- Abstain from alcohol and caffeinated beverages.
- Drink plenty of clear fluids; around 10 to 12 glasses a day. Drinks with electrolytes, such as Gatorade, are a good choice.
- Record your weight twice daily and the number of times you urinate. Contact your doctor if you note a 5lb weight gain in 24 hours or a reduction in the frequency of urination by approximately 50%.
- Be aware of bodily changes and call the doctor with any pelvic pain or increasing symptoms.

CLINICAL TREATMENT OF OHSS:

- Cancellation of embryo transfer until a later date
- Prescription anti-nausea medication
- Drain excess fluid from abdomen with a needle puncture similar to an egg retrieval
- Administer intravenous fluids

Prevention is better than cure and your doctor should monitor the stimulation phase of the IVF cycle carefully. Ultrasounds and blood tests measure the ovaries response to the stimulation drugs, the dose will be lowered if the estrogen levels become too high.

Complications from Retrieval

There are very small risks involved with undergoing egg retrieval under anesthesia. These can include damage to internal organs, inadvertent puncture of a small blood vessel and infection. These complications are extremely rare and are not a cause for concern. The chances of a complication from egg retrieval are less than 1 in a 1000. Talk with your doctor for reassurance about the egg retrieval if you are at all worried.

Adnexal Torsion (ovarian twisting)

Adnexal torsion refers to a condition where the ovary twists and cuts off the blood supply. Women undergoing fertility treatments are at slightly higher risk than normal due to the increased size and weight of the ovary. Torsion of the ovary is rare and occurs in only 0.2% of all cycles.

Ovarian torsion may cause severe pain and tenderness in the lower abdomen along with nausea and vomiting. Treatment requires surgery to untwist the ovary and, in very severe cases, the ovary may need to be removed. Call your doctor if you are having pain, early intervention is best.

You can reduce your chances of this happening by limiting exercise during the stimulation phase of the IVF cycle until the ovaries return to their original size.

Ovarian Cancer

Many have worried that the use of fertility drugs can put a woman at increased risk of cancer; particularly cancer of the reproductive organs. Women who suffer from infertility have an inherently higher risk of these types of cancers so it is difficult to determine whether fertility drugs are causing the increased occurrence or whether it is due to pre-existing conditions associated with infertility.

Several studies have shown women who take fertility drugs do have an increased risk of some types of cancer, but when the analysis takes into account the increased risk due to infertility per se, the evidence does not support an increased risk due to fertility drugs. In other words, there is no cause and effect. Women at greatest risk of ovarian cancer are those who are infertile and who do not get pregnant. Even in this case, the risk is very small; less than 1%.

More research is required to examine the long-term relationship between fertility drugs and cancers of the reproductive system. For now, studies have shown it safe to proceed with fertility medication.

Cancellation of Cycle

At any point during an IVF cycle, treatment may be cancelled. This is usually due to a low response to the stimulation medication, in which case the cycle is cancelled before egg retrieval. If the cycle is cancelled due to a low response, the doctor can sometimes try a different protocol for the next attempt. This may include a higher dose of stimulation medications and/or less suppression medication. If there are several follicles growing, and the sperm quality is good, a cancelled IVF cycle due to low response is often converted to an IUI cycle so all is not lost that month.

Egg retrieval may also be cancelled due to a high response to the stimulation medication (with risk of OHSS), presence of ovarian cysts following Lupron, illness or for personal reasons. Each clinic has their own criteria for when a cycle is cancelled during the stimulation phase or when all the embryos are frozen and the embryo transfer is cancelled. Some physicians will go ahead regardless, if the patient is agreeable and is not at risk of complications. In some instances, the embryo transfer is cancelled following the egg retrieval and the embryos are frozen for future use. Reasons for this include OHSS, fluid accumulation in the uterus, illness or a premature rise in progesterone during the stimulation phase. While this can be quite upsetting to the patient, it is better to freeze the embryos and replace them in optimal conditions.

POSSIBLE REASONS FOR CANCELLATION OF AN IVF CYCLE:

- Low response to stimulation medication
- Over-suppression of the ovaries
- Under-suppression of the ovaries (cysts present)
- High response to stimulation medication
- OHSS
- Poor endometrial lining development
- Premature luteinization (a rise in progesterone too early)
- Premature ovulation
- Fluid in uterine cavity
- Illness
- Psychological reasons

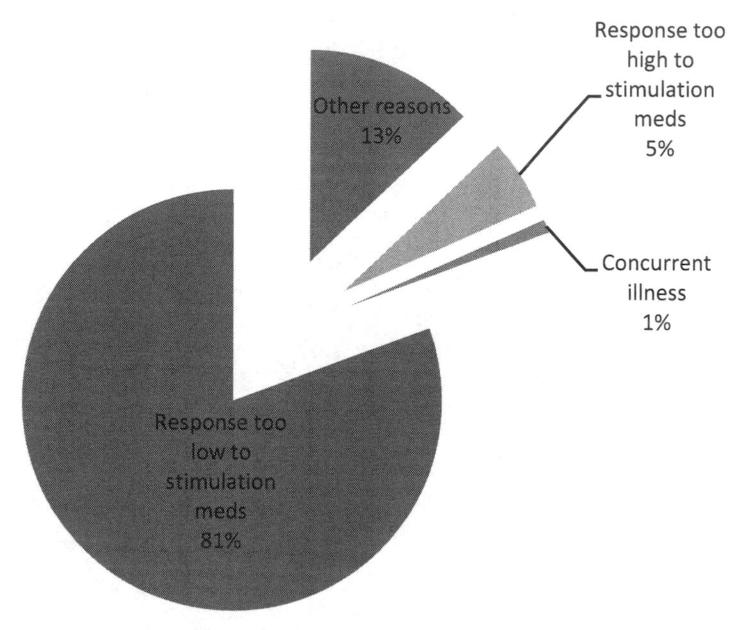

Psychological Effects

Infertility can affect a couple emotionally and psychologically. While it is normal to experience emotional ups and downs during fertility treatments, it is important to recognize when the feelings are severe and you need some extra help.

Most clinics offer emotional as well as physical support and may have specialized staff available for counseling. If you experience prolonged periods of depression or anxiety you may benefit from speaking with a mental health professional. Ask your clinic for a referral if you think this would help you through the process.

Some people get comfort and reassurance from others who are going through a similar experience. Ask your clinic if they have a support group or think about starting one yourself. Try www.ivfconnections.com to connect with other people going through IVF. The organization RESOLVE also has resources to help and support people going through infertility.

Complications of Pregnancy

Infertility and the use of assisted reproductive technology have been shown to increase the risk of pregnancy related complications such as pre-term delivery, lower than average birth weights and cord and placental defects. It is not known whether the increased risk is due to the inherent infertility of the woman or the fertility treatments used to achieve pregnancy. Many large-scale studies have compared babies born through IVF and babies conceived naturally and the majority concluded that the techniques used for fertility treatments are safe and that the majority of complications occur due to multiple gestations or pre-existing medical problems.

CORD AND PLACENTAL ABNORMALITIES

Placental defects have been shown to have a higher occurrence in IVF pregnancies than naturally conceived pregnancies. These include placenta previa, vasa previa and cord defects including velamentous/marginal insertion of the cord.

These conditions can be very dangerous to both mother and child particularly late in the pregnancy. It is essential to get good quality prenatal care. Inform your obstetrician that you are pregnant as a result of fertility treatments and that your baby might have a higher chance of compilations during birth than naturally conceived babies. He can reassure you that all is fine with a simple ultrasound. Don't forget there is more chance of everything being perfectly normal than not, but it is always better to be on the safe side.

IMPORTANT

Tell your OB doctor that this is an IVF pregnancy and that the baby is at higher risk of placental and cord abnormalities. Ask for a color Doppler ultrasound scan to ensure the cord insertion and placental development is normal. Ask for this additional support to ensure your baby is not at risk during delivery. This is very important.

ECTOPIC PREGNANCY

Ectopic pregnancy occurs when an embryo implants and grows outside the uterus, this is usually in the fallopian tube but can be on the cervix, ovary or in the abdomen. Compared with natural conceptions, the risk of ectopic pregnancy is reported to be slightly higher with IVF pregnancies. Ectopic pregnancy usually starts with a slow rising, low level of hCG early in the pregnancy. This is a warning sign and your doctor should monitor you carefully.

It is not yet known why IVF would increase the chances of ectopic pregnancy but women may be at higher risk if they have tubal factor infertility.

Unfortunately, ectopic pregnancy cannot be prevented but early intervention can reduce complications. Women with a history of ectopic pregnancy are at a higher risk of having another one. For these women, it may be helpful to place the embryos lower in the uterus during the embryo transfer and in a smaller volume of fluid.

Ectopic pregnancy is diagnosed through ultrasound and can be treated medically with drugs or, in severe cases, with surgery.

MULTIPLE PREGNANCY

The risk of a multiple pregnancy is usually determined by the age of the woman or egg donor. The number of embryos transferred is adjusted for each individual to minimize the risk. When a woman gets towards her late 30's and early 40's we compensate for the genetic changes of the eggs (and lower success rates) by transferring more embryos, but anytime multiple embryos are transferred there is a chance they will all implant. The number of embryos to transfer is a balancing act between the numbers required to achieve a pregnancy while reducing the chance of a multiple pregnancy.

Many fertility treatments result in multiple pregnancies; in fact, more than 30% of IVF pregnancies are twins or higher order multiple gestations. These pregnancies are at increased risk of maternal and fetal complications and can place an enormous strain on the parents.

The maternal risks associated with multiple pregnancies include but are not limited to:

- Miscarriage
- Bleeding
- High blood pressure
- Pre-eclampsia
- Diabetes
- Anaemia

There are several ways to reduce the chance of a high order multiple pregnancy (triplets or more). These include limiting the number of embryos transferred and using blastocyst culture to enhance selection of the most viable embryos. In most age groups, the transfer of more than two embryos does not increase the pregnancy rates but does increase the rate of a multiple pregnancy. Ask yourself, what is the chance that the best two embryos chosen for transfer won't implant but the third one will? It is likely that when the embryologist chooses the best two embryos, this will include the embryos most likely to implant. This is particularly true for women under age 35 or those with good quality blastocysts. With more advanced freezing methods, it is possible to preserve good quality embryos for future use, even if only a single good quality embryo remains. Therefore, embryos will not be wasted if you do not transfer them.

It is better to be conservative with the number transferred and freeze the good ones for future use than face the dilemmas a triplet pregnancy presents.

Options available to a couple if a high order multiple pregnancy is detected include:

- Continue with the pregnancy under the care of a high risk OB.
- A "wait and see" approach. Many pregnancies spontaneously reduce.
- Selective reduction.

In 10–30% of multiple pregnancies one or more of the fetuses will be spontaneously absorbed and will no longer appear viable on ultrasound. This is sometimes called "vanishing twin." The chance of a pregnancy naturally reducing increases with a woman's age and usually occurs before 12 weeks of pregnancy.

Selective reduction is a procedure usually performed when there are three gestations or more. It involves ultrasound-guided injection of a chemical substance into one of the fetuses that stops the heartbeat. This increases the likelihood that the remaining pregnancy will continue and is commonly performed between 10 and 12 weeks of pregnancy.

A high risk OB performs this procedure and only does so after a full evaluation of each fetus to determine if any have an unusual or abnormal growth pattern or indicators of chromosome abnormalities. Selective reduction carries a risk of 3–4% of losing the whole pregnancy and the risk level of the pregnancy remains somewhat high for the remainder of the pregnancy following the procedure.

Many people feel that they would be unable to reduce a pregnancy if faced with triplets or more for social and ethical reasons. It can be a devastating experience to reduce a much-wanted pregnancy after seeing the ultrasound scans for several weeks. Doctors do not typically view triplet pregnancies as successful outcomes of an IVF cycle. For this reason, it is important to work with your doctor to determine the right number of embryos to transfer for you and talk with your partner about how you both feel about selective reduction before treatment begins.

PRETERM DELIVERY AND LOW BIRTH WEIGHT

Several studies have linked IVF babies with pre-term delivery and low birth weight. Most of these low birth weight babies are a result of multiple gestation, but single gestation pregnancies have also been shown to have a higher prevalence. Single babies born through IVF tend to be born slightly earlier than naturally conceived babies (39.1 weeks compared with 39.5 weeks). IVF twins are not born any earlier than naturally conceived twins, however, all twins are usually born at around 35 weeks of gestation. Triplets are born, on average, at 32 weeks gestation and face the most risks due to their low birth weights and underdevelopment.

Multiple pregnancies are at an increased risk of fetal respiratory distress, neurological, growth and gastrointestinal complications.

Fetal complications associated with prematurity can be serious and lifelong. The risk of having a child with cerebral palsy, for example, is eight times higher for a twin pregnancy and 47 times higher for triplets compared with a singleton pregnancy. Stillbirth rates and neonatal death rates are higher with multiple pregnancies compared to singletons. For singleton pregnancies the rate is less than 1%, for twins 4.7% and for triplets 8.3%.

The list of complications for babies born too early is long and heartbreaking. This is why reproductive medicine providers are working hard to reduce the chance of this happening. With advances in embryo culture techniques and the ability to choose the very best embryos for transfer, we are working towards the goal of reducing multiple pregnancy rates while maintaining a high rate of successful outcomes.

Your physician will discuss with you the appropriate number of embryos to transfer, taking into account your age, the quality of the embryos and other medical and personal factors. A couple's feelings on selective reduction may also be taken into account when deciding how many to transfer.

RISK TO OFFSPRING

The physical, social and intellectual development of IVF children appears to be very reassuring according to studies carried out so far, with the majority of the children doing as well as their peers. Most developmental delays observed in IVF children arise from problems associated with prematurity and low birth weights as a consequence of multiple pregnancy, and are not believed to be due to the fertility treatments themselves.

IVF does slightly increase the chance of a child being born with a birth defect. The overall risk of a birth defect in the general population is 2–3% per live birth. In IVF babies the risk is increased to 3–4% and is predominantly seen in male children. These birth defects include imprinting disorders, such as Beckwith Wiedemann syndrome, which are rare genetic disorders.

IMPORTANT

Despite these risks, the chances of having a normal, healthy child through assisted reproductive technology are, overall EXTREMELY high!

Ethical and Religious Considerations

Infertility treatments can raise ethical and religious concerns for some people.

IVF involves the creation of human embryos outside the body and can involve the production of excess embryos as well as high order multiple pregnancies. If you feel concerned about these issues, talk with your doctor and trusted members of your community who will be able to guide and advise you. A doctor may recommend freezing eggs rather than embryos or may do a low dose stimulation to reduce the number of eggs collected, thereby reducing the number of embryos created. Although these approaches may reduce your chance of success, it is important to stay true to your convictions and feel comfortable with the treatment process.

8

Case Studies

In this chapter we look at several real IVF cycles; the embryos created and transferred and the outcome of the treatment.

You will notice that not all the embryos grow to blastocyst by day five of culture. If there are no blastocysts available for transfer on day five, it is possible to transfer morula stage embryos, rather than waiting until day six, which is also an option. This is not necessarily a cause for concern since it can take an embryo up to six days to form a blastocyst. Many people become pregnant following the transfer of morula stage embryos. Also, you will see that some people become pregnant even if they only have 2 or 3 embryos in total.

Remember, it only takes one good quality embryo!

Case Study One

Age	36
Diagnosis	Male factor and diminished ovarian reserve
Stimulation protocol	Flare
Number of eggs	8
Number fertilized	6
Number transferred	2 x Day 5 blastocysts
Number frozen	2
Outcome	Pregnant with 1 baby

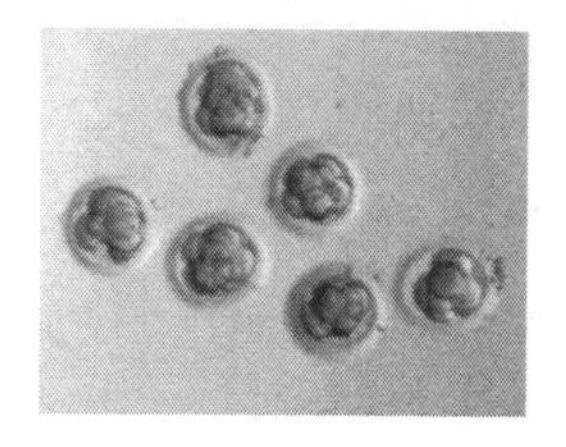

Day 3

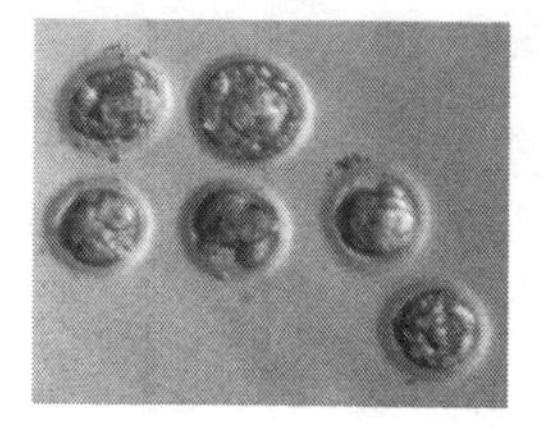

Day 5

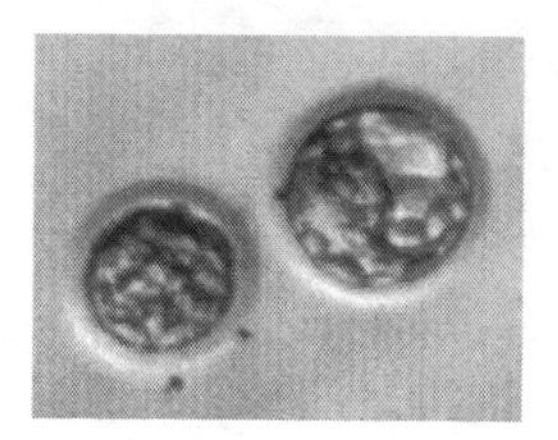

Transferred embryos

IN THE PATIENT'S OWN WORDS

"Having been through the IVF process twice, the process can be just as exciting, scary and overwhelming as the first go around. Having three years between each of the IVF cycles, and the joy of a successful first attempt, certainly proved there was much of the process we had forgotten but also allowed us to experience the advances and process changes which had taken course over three years time. By far, the most overwhelming part of the process is the extensive medications from pills to capsules to injections and trigger shots...all having critical timing impacts and specific roles on our already anxiety filled bodies. I must say that with this second IVF round I am even more confident in a successful outcome based on the medical advancements, research and experience of the Doctors, Nurses and of course, our Embryologist."

Case Study Two

DONOR EGG IVF

Age	40
Donor age	26
Diagnosis	Diminished ovarian reserve
Stimulation protocol (donor)	Birth control pill/Lupron overlap
Number of eggs	6
Number fertilized	4
Number transferred	Elective single embryo, day 5 blastocyst
Number frozen	3
Outcome	Pregnant with 1 baby

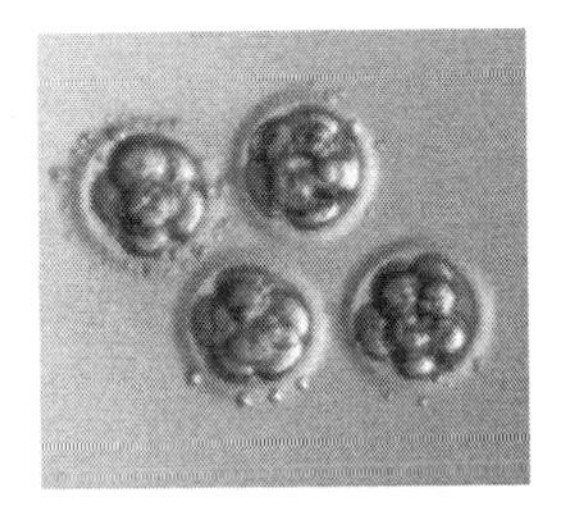

Day 3

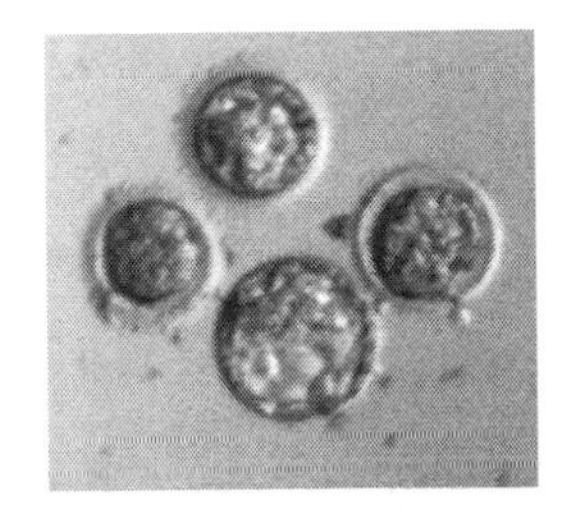

Day 5

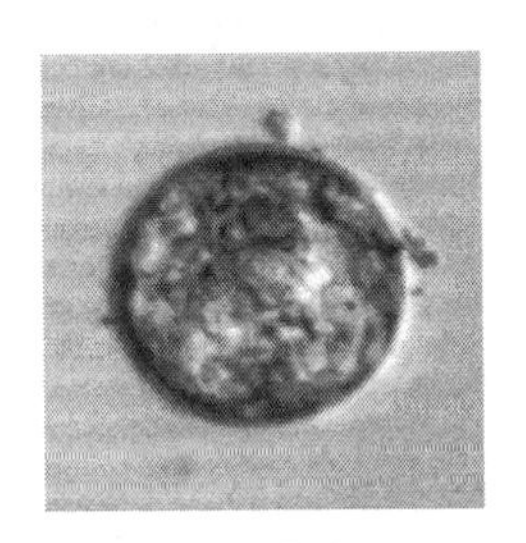

Transferred embryo

IN THE PATIENT'S OWN WORDS

"I was 39 when I was diagnosed was premature ovarian failure and unable to make my own eggs any longer. I started researching IVF clinics that did donor egg cycles and looked for a clinic with top SART statistics and found Oregon Reproductive Medicine.

"Having never been through an IVF cycle I had a lot of learning to do and had to find a donor as well. Once I found her, I patiently waited during the cycle to hear her antral follicle count and how she stimulated. She had donated 6 months prior at a different clinic, and had 26 eggs retrieved resulting in 8 embryos and a singleton pregnancy (the recipient transferred 1 and froze 7). Expecting similar numbers for my own cycle I was happy to hear that she had 20+ follicles on one of her ultrasounds prior to egg retrieval. I eagerly awaited the results of the number of eggs retrieved and got the call that only 6 were retrieved. Somehow, she started to ovulate prior to the retrieval. That was a very difficult time for me as I never expected to only get 6 eggs! (Especially from a healthy 26 year old.) I was petrified that none would be mature enough to use and the entire cycle would be a bust. Luckily 5 of those were mature and ICSI was performed on all 5 resulting in 4 that fertilized and made it to day five and six (3 were frozen). I transferred 1 and am currently pregnant.

"The hardest part of this process is letting go and trusting in it. The unknown was very stressful for me, but all in all it was a very positive experience and I couldn't be happier to finally be pregnant! My opinion is that the success rates of the clinic you select are essential–a clinic with high stats is a result of both the doctors and the embryologists/lab. I had many choices with lower success rates locally, but felt it important to go to the best clinic I could and as a result had to

travel for both the initial exam and retrieval. In the end it really wasn't such a big deal to use a clinic out of state and I would do it all over again."

Case Study Three

Age	41
Diagnosis	Severe Male Factor and advancing maternal age
Stimulation protocol	Birth control pill/Lupron overlap
Number of eggs	29
Number fertilized	9
Number transferred	3 x day 5 blastocysts
Number frozen	1
Outcome	Pregnant with 1 baby

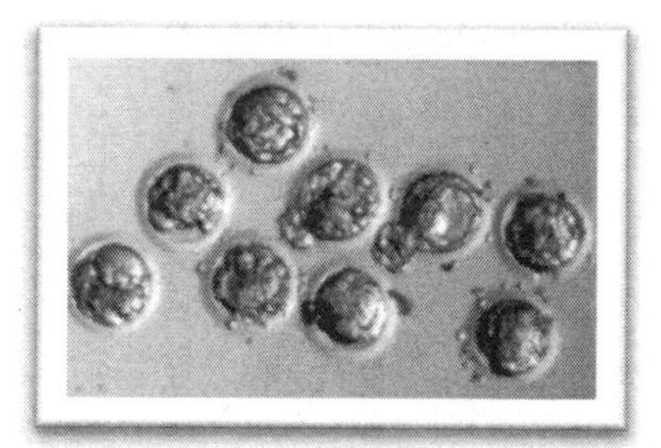

Day 5

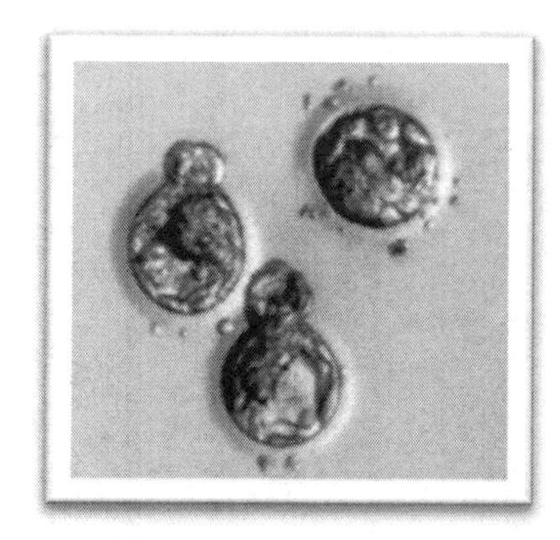

Transferred embryos

IN THE PATIENT'S OWN WORDS

"My husband and I each have our own fertility issue. He is 33 years of age and had zero sperm in his ejaculate. In order for us to pursue a pregnancy that shares our DNA it was necessary for him to go through several sperm retrievals. These procedures required the removal of portions of his testicle. Our Embryologist then teased the tissue apart in the hope of finding pockets of sperm. His second surgery resulted in sperm quantities in the 10's. His third gave us 1000's! We now have 3

vials in storage. This type of sperm is called mesa-sperm. It contains DNA, but may not fertilize an egg as consistently as ejaculated sperm.

"My fertility challenge is that I am 41. I have been told this is advanced when it comes to playing the fertility game. Therefore, I was thrilled to learn that 29 eggs were retrieved, 20 of which were mature. Unfortunately, only 9 fertilized due to the use of the mesa-sperm. On day 5 we had four promising embryos. We transferred three and froze one. Currently I am 8 weeks pregnant with a singleton. We are very happy, but also guarded. We understand that my age leaves our pregnancy at a greater risk for miscarriage and chromosomal abnormalities."

Case Study Four

Age	39
Diagnosis	Diminished Ovarian Reserve, High FSH
Stimulation protocol	Letrozole (mini-IVF)
Number of eggs	2
Number fertilized	2
Number transferred	2 x Morula stage on day 5
Number frozen	0
Outcome	Pregnant with 1 baby

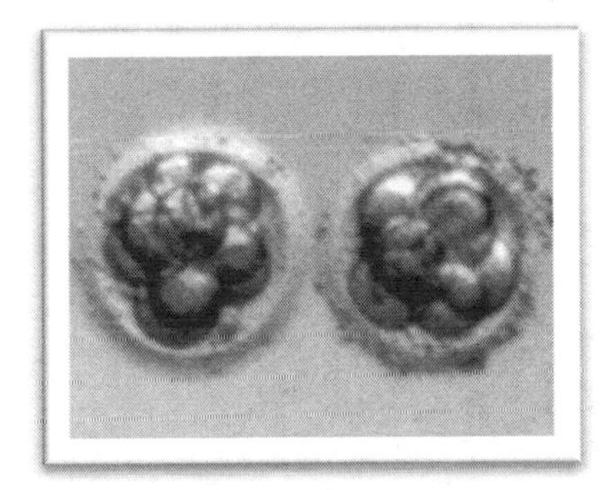

Day 3

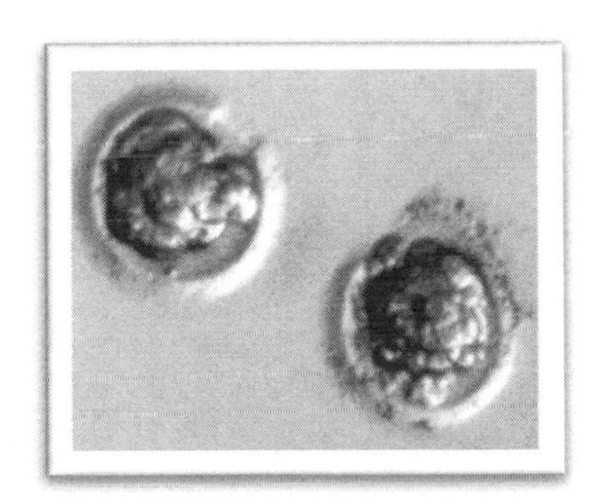

Day 5, transferred embryos

EMBRYOLOGIST NOTES

This couple came to the clinic in 2007 following referrals from several friends. They had been trying to get pregnant for almost four years and had not done any prior fertility treatments. A hysterosalpingogram revealed a normal uterus and open fallopian tubes. A Clomiphene Challenge Test (CCT) revealed high FSH levels, indicating diminished ovarian reserve. Semen analysis was normal. Several IUI cycles were carried out without success.

Two years later, this couple returned to the clinic wishing to undergo IVF. Standard testing was carried out, including a repeat Clomiphene Challenge Test, sonohisterogram and semen analysis. The sonohisterogram revealed some uterine cavity abnormalities, which were corrected during a hysteroscopy.

Due to the female partner's raised FSH levels, it was decided to proceed with a mini-IVF cycle. This involves stimulating the ovaries with Clomid or Letrozole prior to the egg retrieval. It was predicted that the chance of conception was similar for a mini-IVF cycle and a fully stimulated IVF cycle.

During the Letrozole-IVF cycle, two eggs were retrieved and both fertilized using conventional insemination. The embryo quality was exceptionally good on day three; both embryos were at the eight-cell stage with top grades. The embryos were hatched and transferred on day 5 at the morula stage of development. A jubilant couple discovered they were pregnant nine days later, and has a healthy ongoing pregnancy.

Case Study Five

Age	35
Diagnosis	Unexplained Infertility
Stimulation protocol	Birth control pill/Lupron overlap
Number of eggs	15
Number fertilized	5 fertilized with IVF, 4 fertilized with ICSI

Number transferred	2 x Day 5 embryos, early blastocyst
Number frozen	5
Outcome	Not pregnant

IVF Embryos

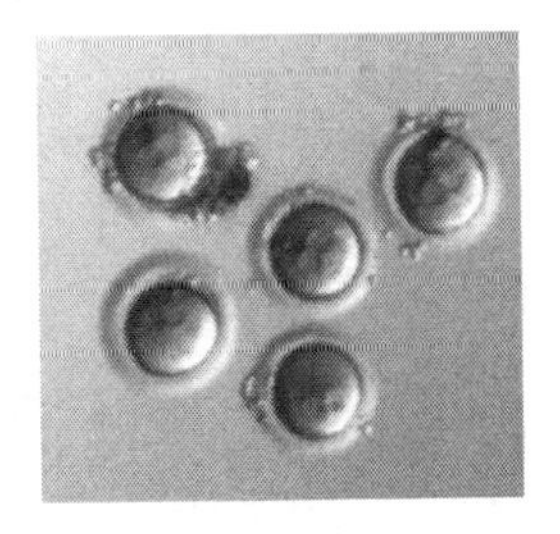
Day 1

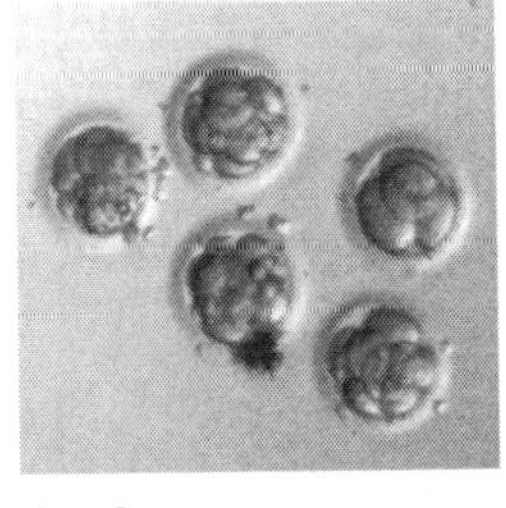
Day 3

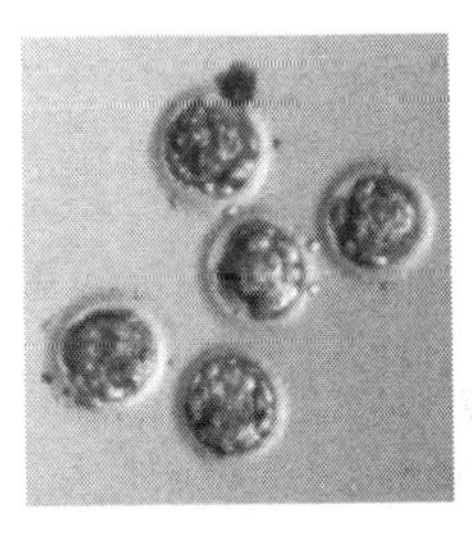
Day 5

ICSI Embryos

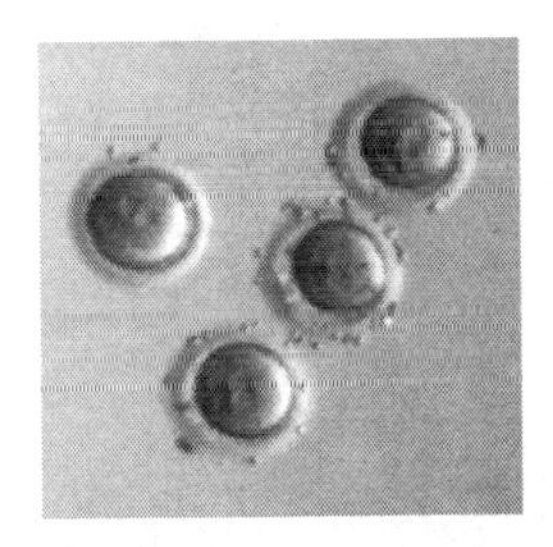
Day 1

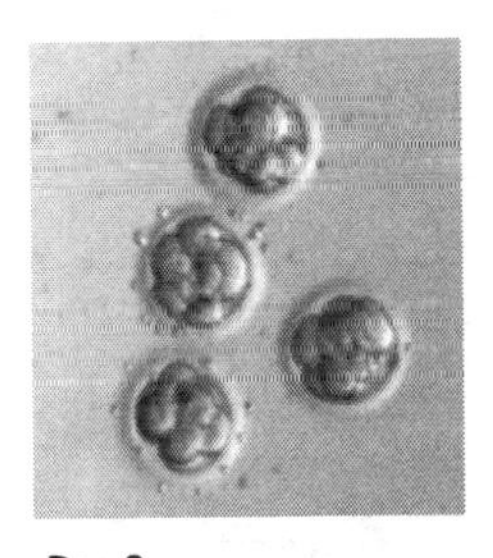
Day 3

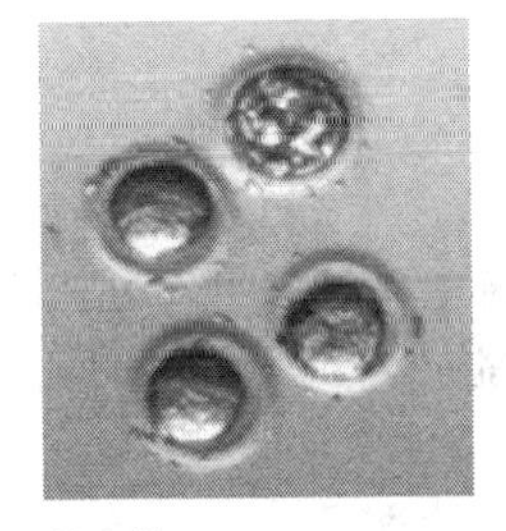
Day 5

EMBRYOLOGISTS NOTES

The wife's gynecologist referred this couple to us after attempting to get pregnant for three years. Prior basic fertility testing by the gynecologist failed to find a diagnosis.

Further testing was carried out at the IVF clinic and included a hysterosalpingogram, ovarian reserve measurement and semen analysis. All test results came back within the normal range. The couple chose to do three cycles of Clomid IUI, which did not result in pregnancy. For this reason, it was decided to proceed with a cycle of IVF. Due to the lack of a diagnosis, half of the eggs were injected with sperm (ICSI) and half were inseminated. The successful natural fertilization of the eggs proved that the underlying problem is not due to an egg-sperm binding issue. The best two embryos were transferred and five embryos were frozen. Unfortunately, the pregnancy test was negative.

We are hopeful of a successful cycle with the frozen embryos in the near future.

Case Study Six

Age	28
Diagnosis	Polycystic Ovarian Syndrome
Stimulation protocol	Birth control pill/Lupron overlap
Number of eggs	13
Number fertilized	6
Number transferred	2 x Day 5 embryos; early blastocysts
Number frozen	0
Outcome	Not Pregnant

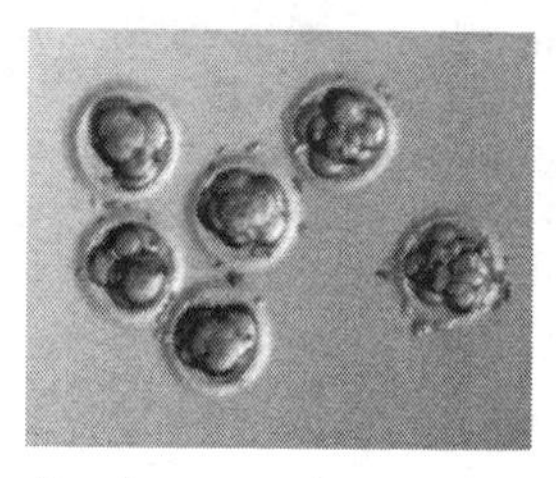

Day 3

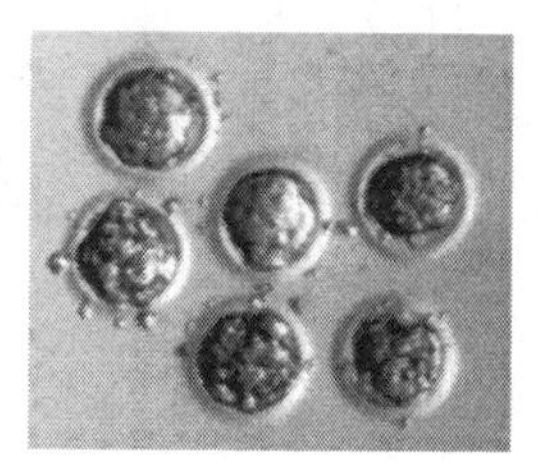

Day 5

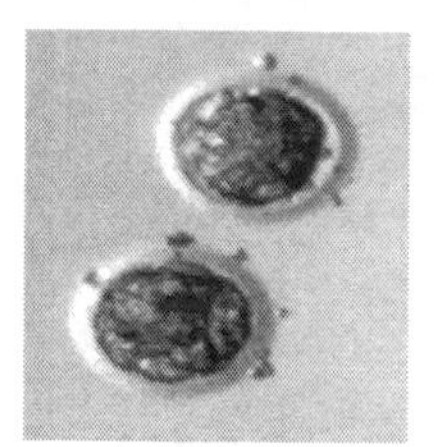

Transferred embryos

EMBRYOLOGIST'S NOTES

The wife's gynecologist referred this couple to us after attempting to get pregnant for 18 months. The female partner had a long history of Polycystic Ovarian Syndrome (PCOS), which was evident by irregular menstrual periods, failure to detect an LH surge on ovulation detection kits and other symptoms.

Upon visiting the clinic, several blood tests were conducted including FSH, LH and estrogen levels. An ultrasound was performed which confirmed the presence of polycystic ovaries.

A hysterosalpingogram showed a normal uterus and fallopian tubes and semen analysis was within normal limits.

Because many women with PCOS respond well to ovulation induction, the first treatment was Clomid therapy with IUI. Following two unsuccessful cycles of Clomid IUI it was decided to progress to IVF.

On a low dose stimulation protocol, 13 eggs were retrieved and six fertilized with ICSI. (ICSI was done due to the couple's relatively long history of primary infertility). Although the embryos were not fully expanded blastocysts on day five, two good quality early blastocysts were transferred. Unfortunately they did not conceive during this IVF cycle.

It is recommended that they try another cycle of IVF in the future following management of some of the symptoms of PCOS. We are very hopeful of a successful pregnancy in the future for this couple, especially given the female partner's young age.

9

Why IVF Fails

Not all patients conceive with their first IVF procedure, which can be a devastating blow to a couple both emotionally and financially. Couples undergoing IVF may need more than one try because of the cumulative nature of conception. Although the pregnancy rates for subsequent cycles are slightly lower than the first cycle (if the first cycle was negative) many issues seen in a first IVF cycle can be addressed and improved upon in subsequent cycles. IVF is therefore both a diagnostic and therapeutic procedure and helps to diagnose problems of fertilization and embryo development that cannot be assessed by any other method.

After a failed cycle the couple usually meets with their doctor to discuss the details of the treatment and their options, which likely include either a frozen embryo transfer or another fresh IVF cycle.

During this meeting, the doctor will review the ovarian stimulation process, any issues with the egg retrieval, egg quality and/or quantity, embryo grades and any problems with the transfer procedure. Often there will be issues in one or more of these areas. At times the doctor can find no reason why the IVF cycle failed and may put it down to "bad luck." This will probably not make you feel any better, but rest assured that you still have a chance of success with a more tailored approach.

Fertility involves many factors and it is not always easy to identify where the problem originated after a negative outcome. IVF can fail for several reasons and can be broken into the following categories, which we will look at in greater detail.

1. Age and ovarian reserve
2. Egg and embryo quality

3. Medical diagnosis
4. Uterine issues
5. Immune and blood clotting factors
6. Sperm Factors
7. Stimulation protocols
8. Embryo transfer
9. Clinic and laboratory
10. Errors with medication/instructions

Biological Causes

AGE AND OVARIAN RESERVE

The most important factor when predicting the outcome of IVF is the age of the woman.

Women are born with a finite number of eggs. As a woman ages, the quality of her eggs declines and chances of her having genetically abnormal embryo's increase. Genetically abnormal embryos usually do not implant at all or are lost shortly after implantation. A woman's age not only affects her chances of conception but also affects the risk of miscarriage. Most miscarriages are due to genetic abnormalities in the fetus.

As well as age, a high FSH level or abnormal Clomiphene challenge test can also predict reduced egg quality. Age and ovarian reserve are both taken into account when predicting the likelihood of success from an IVF cycle.

Medicine is unable to reverse the aging process in eggs. The best way to maximize the chance of success is to get as many eggs as possible in the hope that the "good eggs" are included. In general, the more eggs we get the better chance we have of a successful outcome in all age groups.

Treatment options for women with poor quality eggs are limited. Sometimes changing the stimulation protocol or changing the laboratory conditions in which the embryos grow can be helpful.

For patients where egg quality is clearly the problem, egg donation may be an option to achieve a successful pregnancy.

The following graph shows the decline in a woman's fertility after the age of 35.

PREGNANCY AND LIVE BIRTH RATES ACCORDING TO AGE (SART)

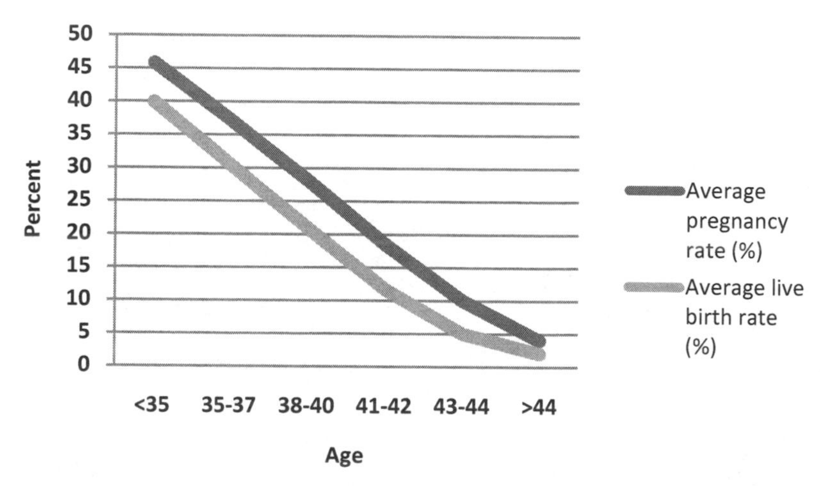

In a small number of patients who have recurrent implantation failure or long standing unexplained infertility, one of the partners may carry a chromosomal imbalance such as a translocation, where parts of the chromosome are switched or crossed over.

This can be tested for on both partners by doing a karyotype; a blood test that analyzes the chromosomes in a person's cells. If a chromosomal abnormality is detected in the parent, then pre-implantation genetic screening (PGS) is used to identify the embryos with normal genetics, thereby increasing the chances of success.

Even if both partners have normal karyotypes, 30–90% (depending on maternal age) of the embryos created in the lab have been shown to be genetically abnormal, and not all genetically normal embryos implant. Improvements in pregnancy rates are made possible by growing the embryos to the blastocyst stage to exclude the "weak" ones and by placing multiple embryos into the uterus during one cycle to maximize the chances that one or two will implant.

EGG AND EMBRYO FACTORS

Most fertility specialists feel that over 95% of IVF implantation failures are due to arrest of the embryo development. This is usually due to chromosome or genetic abnormalities that do not allow normal development; the embryos aren't able to make the correct proteins and therefore do not attach to the uterine lining. Problems with the egg can manifest clinically as diminished ovarian reserve or premature ovarian failure and are predicted by maternal age.

Assisted hatching may improve pregnancy rates by allowing the embryo to easily leave its shell. If the embryo got "stuck" inside the shell at the blastocyst stage, it would not be able to implant and a pregnancy would not follow. Women who have a less than favorable prognosis for pregnancy usually have their embryos hatched prior to embryo transfer. Following a failed cycle if embryo implantation did not occur, some clinics will perform assisted hatching during subsequent cycles to ensure the embryos have every chance to implant.

MEDICAL DIAGNOSIS

The chance of an IVF cycle being successful can be affected by the diagnosis you are given. Some problems are easier to "fix" than others. For example, a young woman with blocked fallopian tubes has a higher chance of success than an older woman with diminished ovarian reserve.

Success rates are similar for different diagnoses. The lowest rates are for women with a low reserve of eggs. These women usually have fewer eggs collected and tend to be slightly older. Women with polycystic ovaries (PCO) may have a higher number of chromosomally abnormal eggs due to raised androgen levels in the ovaries; however, this is partially compensated for with the collection of high numbers of eggs. The good news is that the chance of getting pregnant for women with PCO using fertility treatments is very good, especially for women under the age of 35.

A good clinic will individualize your treatment based on your diagnosis and other factors such as age and ovarian reserve to give you the best chance of success.

The following graph shows success rates by diagnosis from data collected from all the IVF clinics who participate in the SART program. These are average percentages and vary from clinic to clinic.

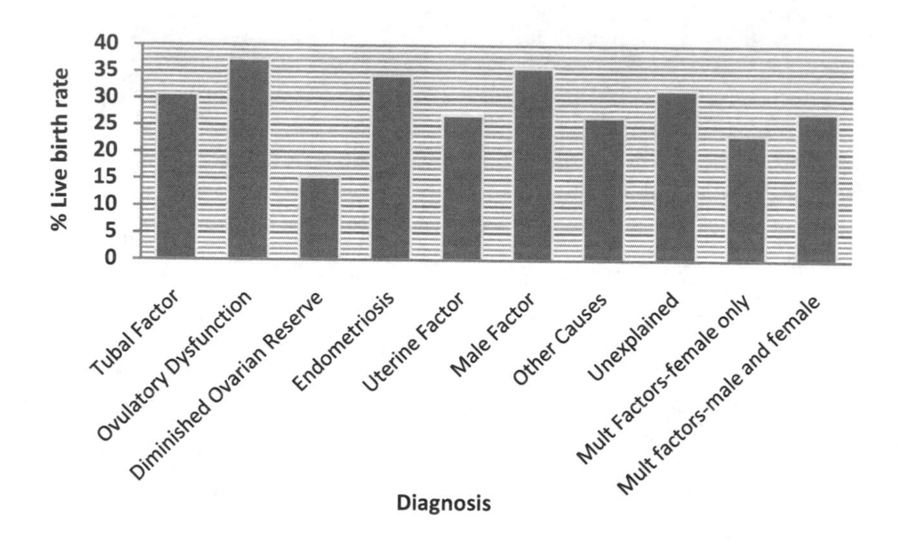

UTERINE ISSUES

It is thought that the uterus is receptive to embryo implantation in the majority of women regardless of age. This is confirmed by the high pregnancy rates seen when older women use donor eggs. However, for some women implantation failure may be due to the inability of the uterine lining to support a pregnancy.

Several studies have shown that implantation and pregnancy rates are lower for donor egg recipients over the age of 45. This is due to unknown biochemical and/or molecular aberrations of the endometrium. The age related decrease is partly correctable by increasing the dose of progesterone supplemented to the recipient during the cycle.

Inflammation due to hydrosalpinx or endometriosis may also play a part in reduced uterine receptivity. Pregnancy rates are higher for women with endometriosis following laser surgery or 3 months of Lupron treatment (to suppress inflammation) compared with women who have no treatment.

Women who use a gestational carrier (GC) usually do so because they have been diagnosed with a uterine abnormality that is beyond surgical repair. Data from SART shows that by using a GC, pregnancy rates were improved for most of the age groups. The biggest surprise was in women over the age of 44 (using their own eggs). The pregnancy rate with a GC was 11.1% compared with a pregnancy rate of 3.1% without a GC. This confirms an age related decline in uterine receptivity with age.

The following graph shows pregnancy rates by age of patients using their own eggs with and without a GC (SART). The age reflects the age of the patient and not the GC.

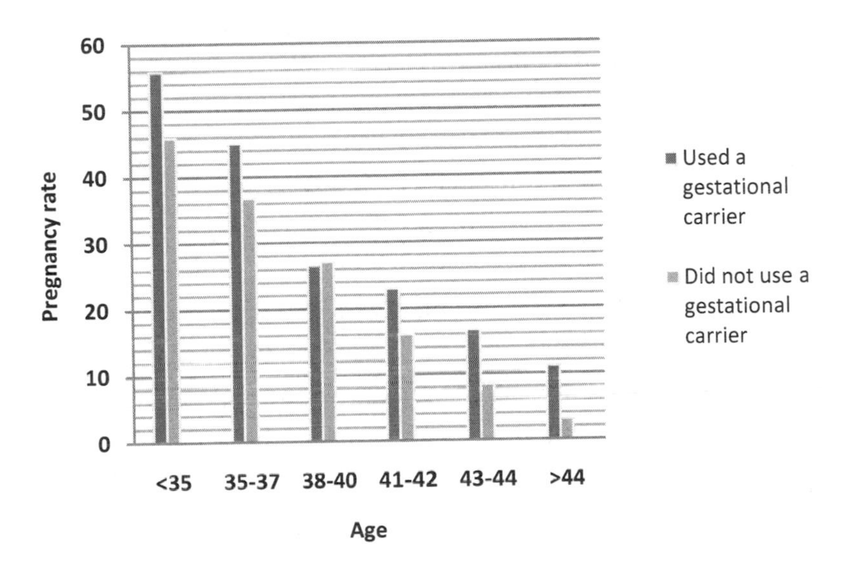

Reasons for a reduction in uterine receptivity can include:

- Anatomical abnormalities (a repeat hysteroscopy may be necessary)
- Lack of expression of beta 3 integrins (or other adhesion molecules)
- Immune issues
- Past or present uterine infection
- Hydrosalpinx or endometriosis causing inflammation of the uterine lining
- Decreased blood flow within the endometrium
- Mutations in some of the genes coding for the progesterone receptor

Some of these conditions can be improved by surgery, changing stimulation protocols, increasing the dose of progesterone, acupuncture or by taking medication. Some women may only be able to have a child using a gestational carrier. It may take several failed treatment cycles with good quality embryos to realize that uterine receptivity is an issue; it is one of the least likely reasons for treatment failure and quite difficult to test for. For this reason, implantation failure is usually diagnosed as unexplained infertility.

IMMUNE AND CLOTTING FACTORS

There is conflicting evidence surrounding the role of the immune system in infertility as well as the tests and treatments used to diagnose and treat autoimmune disorders. At present, there is no agreement in the medical community about what tests should be performed or how to treat an abnormal result.

Despite this, a large number of immunological tests are available in an attempt to identify patients with immune disorders and a variety of treatments are advocated.

Tests include the measurement of:

- Anti-phospholipid antibodies
- Antinuclear antibodies
- Antibodies to alpha-2 glycoprotein 1
- Anti-thyroid antibodies
- Natural killer cells

There is evidence that women with abnormal results from immune testing are at higher risk of having a miscarriage. There is, however, no strong evidence to show that immune issues are the cause of infertility or implantation failure.

If a woman is found to have raised levels of immune cells her doctor may choose to treat these issues during an IVF cycle.

Treatments include:

- Baby aspirin
- Heparin
- Corticosteroids
- Intravenous immunoglobulins (ivIg).

Some studies have shown these treatments improve pregnancy rates whereas others have shown them to have no effect. It is difficult to know whether a successful outcome was achieved because of these additional therapies, or whether it would have happened regardless. Sometimes people get pregnant after a second or third IVF cycle even without making any changes. Some naturally fertile women have been shown to have a high number of immune cells in their circulation and endometrium, disproving to some that immunity plays a role in preventing pregnancy. Each doctor will have their own opinion about the value of immune testing and patients should follow their physician's advice.

Also controversial is the role that blood clotting disorders known as thrombophilias (e.g. MTHFR and Factor V Leiden) play in infertility. These conditions cause an increased chance of abnormal blood clotting and pregnancy complications. Although thrombophilias have been linked with recurrent miscarriages, there is little evidence that they have an impact on implantation and IVF outcomes. At most clinics, immune and thrombophilia testing are not usually recommended unless a woman has a history of recurrent miscarriage (2–3 or more). Ask your doctor if you think you might benefit from this type of testing.

Immune cells in the uterus can actually have a positive effect on embryo implantation as the endometrium becomes rich with immune cells following ovulation. Interactions between the uterus and the embryo are communicated through proteins called cytokines which are secreted by the cells within the uterine lining including immune cells.

If the immune cells don't send the correct signals to the embryo through cytokine secretion, or the cells don't respond to signals from the embryo, then implantation will not occur.

SPERM FACTORS

Sperm quality influences not only rates of fertilization but also subsequent embryo development. Remember, half of the genes come from the father. The male partner may carry a chromosomal abnormality that is responsible for him having a low sperm count and increases the risk of implantation failure and miscarriage.

Standard semen analysis usually tests sperm count, motility and morphology. There are several more advanced tests that can be carried out if the sperm count is low for no apparent reason or if several treatment cycles fail without explanation.

Problems with the sperm not diagnosed by standard parameters of a semen analysis are:

- A balanced translocation
- Y chromosome micro-deletions
- Cystic Fibrosis gene mutations
- DNA fragmentation
- Aneuploidy

These tests can detect problems with the genetic component of the sperm (inherited from the male) and may provide an explanation for why the sperm count is low; embryo quality poor or a pregnancy is not achieved. There are no magic cures to dramatically improve sperm quantity but if a man has a healthy lifestyle and does not drink or smoke he will maximize his chance of having good quality sperm. It takes 12 weeks for sperm to be made in the body so a man should adopt a healthy lifestyle for at least 3 months before a treatment cycle begins.

Several studies have shown increased sperm quality when a man takes vitamin supplements for a prolonged period of time. This is especially true if the supplement contains vitamin C, zinc and folic acid. Eating a balanced, healthy diet with plenty of fresh fruit and vegetables along with good quality proteins and healthy oils has also been shown to improve fertility in men.

Other Causes

STIMULATION PROTOCOLS

There are several ways to stimulate the ovaries to produce many eggs during an IVF cycle; these are called stimulation protocols. An IVF cycle allows the doctor to identify very specific problems that may be unidentified during standard testing. This can include tailoring the stimulation protocol to maximize the number of eggs retrieved.

Following a failed IVF cycle, the doctor will review the stimulation protocol and determine whether changes can be made with a subsequent treatment cycle to improve the chance of success. Trial and error, prior responses and physician experience play a huge role in choosing the optimal protocol for an individual.

EMBRYO TRANSFER

The embryo transfer is technically crucial. A difficult embryo transfer can result in implantation failure despite having good quality embryos. Most often, this is due to trauma to the uterine lining or difficulty placing the embryos in the right place. A mock transfer is usually carried out in the month preceding an IVF cycle to identify if the transfer might pose a problem. If the catheter does not easily pass through the cervix during a mock transfer, the doctor may dilate the cervix before the actual embryo transfer. In addition, a trial pass of the catheter is usually done immediately before the embryo transfer to confirm an easy passage through the cervix. The goal is to have the transfer be as least traumatic as possible to the uterine lining.

This practice, along with ultrasound guidance and relaxation should help the transfer go smoothly. Some doctors are more skilled than others at procedures such as embryo transfers. It helps if your doctor has had plenty of experience and has a good reputation.

If you feel comfortable, make inquiries at the clinic with staff you trust, (a nurse or embryologist) and ask if some of the doctors have higher pregnancy rates per transfer than others at the practice.

CLINIC AND LABORATORY

IMPORTANT

Clinic choice is one of the most critical factors determining success or failure and is one of the few things within your control. Choose wisely.

Once you choose a clinic, the progression of your IVF treatment is mostly in the hands of the doctor and his staff. That is why it's so important to look at success rates and do your homework before you commit to a fertility specialist.

The doctors should be well qualified and come with good recommendations from other patients and your OB/GYN. A fertility doctor should not be afraid to try new approaches to infertility treatments based on each individual and their specific circumstances.

To enable IVF clinics to achieve high pregnancy rates it is important for them to have a state of the art laboratory with highly trained embryologists. They should offer ICSI, assisted hatching, PGS and blastocyst culture (day 5 transfer).

ERRORS WITH MEDICATION OR INSTRUCTIONS

When you start fertility treatments you will be given numerous medications to take. These drugs are taken in different ways, at different times and on different days, so it's no surprise that some people make mistakes with their medication regimen. Some of these drugs are vital to the success of the treatment whereas others are less important. It helps to be very organized and enlist the help of your partner to ensure you follow your regimen. The clinic should give you a calendar to follow with your daily dose and timing of the medication.

If you take the wrong amount, miss a dose or take the wrong medication, call your clinic right away and they will tell you how to correct it. Most of the time it is not a problem and you get right back on track with your regular dose.

Lastly, check your medication when you pick it up from the pharmacy. Read the label and check the name and dose of the drug to ensure it is exactly what was prescribed. Pharmacies have been known to make mistakes and so it is a good idea to always check your medication before you take it.

Try, Try Again

If you have a failed IVF cycle, don't despair; talk with your doctor about how he or she can help you to move forward on your path to parenthood.

- Ask if there were any obvious problems during the cycle
- Get a clear estimate from your doctor on chances of success with a second cycle
- Ask what modifications or changes can be made to improve your chances for another cycle
- If you have frozen embryos, consider the chance of success with these versus the chance of success with fresh eggs (before you get any older)

10

How to Improve Your Chance of Success

Following a failed IVF cycle there is a healing process to go through, take some time to grieve your loss and gather yourself to move forward. It is a good time to consider your options and reflect on the treatment you have received so far. Hopefully you will have the opportunity to pursue further treatments and are optimistic about the future. Depending on your situation, it is usually just a matter of time before you become pregnant. Your doctor should be able to help you understand your options and how further treatments can be improved.

While fertility is mostly out of your control, here are a few ideas to consider which may help to improve your chances of conceiving in the future. Use these as a starting point for discussion with your doctor.

- Change to a clinic with a higher success rate or different protocols
- Change protocols at the same clinic
- Transfer more embryos next time
- Do ICSI if there was a low percentage fertilization with natural insemination
- Try a new technology or "alternative" therapy (e.g. Pre-implantation Genetic Screening, ivIg, Growth Hormone, DHEA or melatonin supplementation, immunosuppressants, etc.)
- Try acupuncture or other complimentary medicine
- Reduce stress
- Make lifestyle changes, such as reducing alcohol, quitting smoking, reducing caffeine and discontinue unnecessary medication

- Take a dietary supplement and eat a healthy, balanced, good quality diet
- Reduce exposure to environmental toxins, cleaning supplies, bleach, other toxic substances such as pesticides
- Reduce/increase body mass index (BMI) to a healthy weight
- Get plenty of sleep

Complementary and Alternative Techniques

Acupuncture and other alternative medicine practices are being used to assist and improve IVF and other fertility treatments. In recent years there has been growing integration of Eastern and Western medicine and it is likely this acceptance within the medical community will continue to increase with time.

ACUPUNCTURE

Acupuncture has been shown in some studies to increase pituitary and ovarian hormone levels, increase blood flow to the uterus and relax the patient. Although there is conflicting evidence as to whether acupuncture can actually increase pregnancy rates, most patients receiving this treatment report that it gives them a sense of well being during a stressful time, making the overall experience a more positive one.

HERBAL MEDICINE

Herbal medicines are a mystery to most people and are usually prescribed by a naturopathic doctor or acupuncturist. Please check with your RE before you take any herbs in case they interfere with other medication you are taking. Several studies have shown that some supplements containing vitamins and other naturally occurring compounds can help with PMS, can lengthen the menstrual cycle and increase fertility. It is difficult to know the exact effect herbal supplements have on fertility as the use of alternative medicine is not always documented by clinical staff. Further studies need to be conducted for a conclusive answer.

The male partner can also benefit from complementary medicine, including acupuncture, and should eat a balanced diet with the recommended amount of vitamin C, zinc and folic acid. These supplements have been shown to be safe and aid in the production of sperm.

DHEA OR MELATONIN SUPPLEMENTATION

DHEA, which stands for Dehyroepiandrosterone, may be a natural fertility booster for women who have poor quality eggs and have been given a low chance of conception. It is a natural steroid produced in the body and serves as a precursor for male and female sex hormones (androgens and estrogens). Natural DHEA levels begin to decrease after age 30.

Several studies have followed women taking a daily DHEA supplement of 50–75mg per day for 1–4 months prior to fertility treatment. The results have shown an overall improvement in ovarian reserve measurements and higher pregnancy rates in this group compared with women who did not take DHEA.

It has been suggested that DHEA may increase the number of eggs collected during an IVF cycle, improve egg quality, increase pregnancy rates and decrease miscarriage rates. This is in addition to giving women an overall improved feeling of wellbeing.

Of course no drug is without side effects; women taking DHEA have reported changes in hair growth, mood and acne due to higher than normal levels of androgens and estrogens in the body. The long term effects of DHEA are not known and it is not recommended for long term use without the supervision of a licensed healthcare professional.

A recent study has found a correlation between melatonin supplementation and improved egg quality. Although this study is yet to be confirmed, it provides further indications that herbal and natural supplements may be the key to improving egg quality and possibly reversing some of the effects of aging we see on ovarian function.

STRESS REDUCTION

Fertility treatments can be very stressful. It is helpful to find a way to ease the stress and take some time for yourself–massage, yoga, reflexology, aromatherapy and gentle exercise might help. Sharing your feelings with your partner, a friend or therapist can also help to ease the burden. People have tried meditation, hypnosis, visualization techniques, journaling and prayer to help them stay optimistic and keep negative feelings in check.

Everyone is different so you should find what works best for you. Having coping mechanisms in place before you start treatment is very important especially if you have experienced anxiety in the past.

Optimizing Natural Conception

Several studies have identified how the chances of conceiving naturally can be maximized. Although these reports are mostly based on couples without infertility, this is still good information for anyone trying to get pregnant. The Practice Committee of the American Society outlines these studies in a report for Reproductive Medicine (ASRM) "Optimizing Natural Fertility" Fertility and Sterility journal, 2008.

TIMING OF INTERCOURSE

During a woman's cycle there are three to six days during which she is fertile. The most fertile days are the two days before ovulation, so this is the optimal time for the couple to have intercourse. It is recommended to have frequent intercourse beginning when the woman's period ends in order to ensure the six-day window of fertility is not missed.

Several studies have shown that a man's sperm does not decline in quality with frequent ejaculation and, in fact, that prolonged abstinence can be deleterious to fertility. The peak in sperm motility is after 0–2 days of abstinence and it is recommended that the couple have intercourse every 1–2 days while trying to conceive. Too frequent intercourse can be quite stressful and exhausting, so the optimal frequency should be left up to the couple to decide.

Some women test the consistency of their cervical mucus to determine the most fertile days of their cycle. Clear, stretchy and slippery type mucus is a very accurate predictor of the fertile phase. This has been shown to be a very good predictor of the chance of conception, even more so than the timing of intercourse. If you are trying to conceive naturally, learn about the changes in your cervical mucus and what to look for. This will ensure that the timing of intercourse is optimal. This can be done in conjunction with basal body temperature monitoring and home ovulation predictor kits.

Lifestyle Factors

Many lifestyle factors can affect fertility. Some things need to be avoided altogether, some moderated and others added. Part of the ASRM report looks at lifestyle factors and determines what, if any, affects your chance of conception. Here are a few of the findings.

ALCOHOL

Many women choose to stop drinking alcohol altogether while attempting to get pregnant which is a good contribution to overall health. Some studies have shown a decrease in fertility in women who have a moderate to high number of alcoholic drinks per week (6–12), while other studies have shown that up to 2 drinks a day has no effect on fertility.

The conclusion from most of the studies of alcohol and its effect on fertility is that a low amount of pre-conception alcohol does not affect fertility but that higher amounts do (especially over 14 drinks per week). To be safe, it is a good idea to cut back on your alcohol intake and, if you feel comfortable, eliminate it altogether.

For people who like to have the occasional glass of wine, don't worry! A survey of pregnant women was carried out to determine their pre-conception alcohol consumption and time it took them to get pregnant. The results showed that moderate consumption of alcohol was not associated with a longer time taken to conceive. In fact, it was shown that those who had a consistent low intake of alcohol (1–2 drinks a week) had a shorter waiting time than non-drinkers!

The only group of women who had a significantly longer waiting time to get pregnant were those who drank more than 14 alcoholic drinks per week. Heavy drinking by the male partner also had a profound effect on the time taken to conceive and was associated with a two-fold increase in waiting time.

The ASRM study concludes that a woman can safely have up to two alcoholic drinks a day while trying to get pregnant and that this will have no adverse effect on fertility. Remember, all alcohol intake should be stopped when you find out you are pregnant.

CAFFEINE

Several studies have been carried out to determine if caffeine intake has a detrimental effect on fertility or pregnancy.

Studies have found that high levels of caffeine (over 5 cups of tea or coffee per day), can increase the risk of miscarriage and reduce fertility, but overall, moderate caffeine intake does not affect fertility or the health of a pregnancy.

It is recommended that a woman have only one to two cups of coffee or other caffeinated beverage a day to ensure fertility and pregnancy are not compromised. Some women prefer to eliminate caffeine altogether to create the healthiest environment possible for conception to occur.

BODY WEIGHT

Being underweight as well as being overweight can affect your chances of getting pregnant. Both of these groups have a four times higher chance of having problems conceiving compared with women with a healthy BMI. For an overweight woman, a weight loss of 5–10% can dramatically improve the chance of getting pregnant. Heavier men also face fertility problems due to perturbed hormone levels.

Weight loss involves a serious commitment involving diet and exercise. It only takes a small change in your weight to make dramatic changes in your health. In fact, a small weight loss may increase natural fertility sufficiently to eliminate the need for medical assistance to get pregnant. Ask your healthcare provider for help finding a weight loss program to suit you.

Fertility issues with obesity	♦ Irregular or infrequent menstrual cycles ♦ Increased risk of infertility ♦ Increased risk during fertility surgery ♦ Increased risk of miscarriage ♦ Decreased success with fertility treatments
Potential pregnancy complications with obesity	♦ Increased risk of high blood pressure ♦ Increased risk of diabetes in pregnancy ♦ Increased risk of birth defects ♦ Increased risk of high birth-weight infant ♦ Increased risk of Cesarean section

Benefits of weight loss	◆ Weight loss of 5–10% can dramatically improve ovulation and pregnancy rates ◆ Improved health including reduces diabetes, high blood pressure and heart disease ◆ Improved self esteem

DIET AND SUPPLEMENTS

A study at Stanford Medical School looked at the effects of a supplement called Fertility Blend on the pregnancy rates of women who were having trouble conceiving.

Fertility Blend contains chasteberry, green tea, vitamins, folate and minerals. The participants in the study were 93 women who had tried unsuccessfully to conceive for between six and 32 months. 53 of the women were given the supplement and 40 women were the control group and did not receive Fertility Blend.

After three months, 14 out of the 53 women (26%) who took the supplement were pregnant, compared to only four of the 40 women (10%) in the placebo group.

The researchers concluded that supplemental nutrients could be a beneficial alternative or complimentary therapy to traditional infertility treatments. The small number of participants, however, limits this study's accuracy, and more research is needed to confirm these results.

It is possible that with additional supporting evidence, herbal supplements will become more widely accepted as a complement to traditional medicine. For now, most clinicians ask that you do **not** take herbal remedies during a medicated treatment cycle, only a recommended prenatal vitamin. Some have been shown to interfere with the way a drug works and can reduce its potency. Always tell your doctor if you are using complimentary therapy as it may be dangerous to your health or have a negative impact on your outcome.

Another study from the UK showed that women who took a multivitamin consistently had a lower chance of developing ovulatory infertility compared with non-users. The researchers estimated that around 20% of all cases of ovulation disorders could be avoided if women take three or more multivitamins per week, particularly those that contain folic acid, iron, vitamins B1, B2 and D, with folic acid being the most important component.

Folic acid supplementation is already recommended for women wishing to become pregnant to prevent neural tube defects. The added effect of decreasing ovulatory infertility is a plus.

After a full review of all the data available, the ASRM concludes that there is little evidence that vegetarian diets, low-fat diets, vitamin enriched diets, and anti-oxidants or herbal remedies improve overall fertility. They did, however, recommend that women attempting pregnancy take a folic acid supplement to decrease the risk of neural tube defects in the fetus.

ENVIRONMENTAL FACTORS AND STRESS

Environmental toxins can reduce fertility. Do what you can to remove toxic materials from your personal and work environment. These can include pesticides, paint fumes, radiation, solvents and chemical cleaners. Eat healthy, natural and organic foods as often as possible. Avoid any type of home improvement that involves the use of building materials and chemicals. Lastly, take a look at your cleaning products and try to use as many natural products as possible. Avoid strong smelling or chemical cleaners, especially sprays, which could be easily inhaled. Look for ones that don't contain harsh solvents, fragrances, chlorine, or ammonia.

Prospective fathers may want to follow many of the same guidelines as the female partner for avoiding toxins. According to the National Institute of Occupational Safety and Health, ongoing exposure to **pesticides, chemical fertilizers, lead, nickel, mercury, chromium, ethylene glycol ethers, petrochemicals, benzene, perchloroethylene, radiation, and other toxins can lower sperm quality as well as quantity, and possibly lead to miscarriage in their partner's pregnancy.**

Extremely high levels of stress have been shown to affect the hormone balance in the body that may affect fertility. If you recognize that you have prolonged or extreme stress, try to postpone reproduction until you can leave that environment or can eliminate the stress. One of the most beneficial reasons for eliminating stress (apart from feeling good) is that it helps people to pursue conception for a longer period of time following unsuccessful cycles. Due to the cumulative nature of conception, the longer you try the more chance you have of getting pregnant. So stress reduction does indirectly increase pregnancy rates!

Many people trying to get pregnant naturally or undergoing fertility treatments feel a certain amount of stress and anxiety. This is perfectly normal and will not affect the outcome. Acupuncture during the preconception period can be very beneficial and has been shown to increase pregnancy rates in some studies.

Summary

Factors that most affect the time taken for a woman to get pregnant are age, obesity, and excessive alcohol intake and smoking. This is also true if the male partner falls into any of these categories. Heavy smokers take twice as long to get pregnant as non-smokers, whether the smoker is the man or the woman.

A woman will have a better chance of conceiving if she and her partner take care of themselves, both physically and mentally, and find time to do the things they enjoy in life. Good nutrition, healthy habits, and exercise are the most important aspects of getting ready for pregnancy.

While trying to conceive:

- Maintain a healthy weight
- Take a multivitamin with folic acid (men too!)
- Don't smoke
- Herbs may help but check with your doctor first
- A low level of alcohol is OK
- A low to moderate level of caffeine is OK
- Take good care of your general health
- Reduce stress and exposure to environmental toxins
- Avoid recreational drugs

Based on study results, couples that lead healthy lifestyles are half as likely to be infertile.

11

Ten Women's Personal IVF Stories

In closing this guide, I felt it best to share with you some personal perspectives and advice on undergoing fertility treatment. Though from varied backgrounds and undergoing a different experience, each of these women all shared the same hope. May your time as a patient be short and your time as a parent be long and rewarding; no matter which path you take to get there.

Michelle

Before we got married, my husband had already undergone extensive fertility treatments (including surgery) in a previous marriage. He was discovered to have a zero sperm count from a blockage in his epididymus. The doctor discovered he produced sperm but they didn't mature properly. My husband underwent surgery and, after that, doctors said everything should work properly. When we married, I already knew the issue and that when we were ready to have children, we'd have to go the IVF route. I was 25 years old.

When it was time to choose a clinic I researched different clinics online. We lived about an hour away from the one we ended up going to, which was the closest clinic with the best reputation and highest success rates. I didn't have to have many tests after we informed the doctors about my husband's medical history; just some of the basic ones to make sure my eggs were ok.

During our IVF cycle I was able to produce 16 eggs, 11 mature and five fertilized into beautiful embryos!!! We decided to transfer two fresh embryos the first cycle. We got ONE beautiful, perfect, healthy baby boy! I couldn't believe it had worked!

We froze the remaining three embryos, knowing we wanted to try for one more. When our baby boy was one year old, we went back and began FET. All three embryos thawed and were beautiful! The embryologist did not recommend we put back all three due to my age (being under 30) and having already had a successful pregnancy. Of course we had gone over this scenario beforehand, but until it happens it's a lot easier to think about options, as opposed to making the decision. We are Christians and that relationship is based upon FAITH! The morning of the transfer, we knew that this was a part of the process that we would not control. We chose, against the advice of the doctors, to transfer all 3 "home" to their Mama! That decision was the best decision we could have made. It resulted in a second pregnancy, with ONE more beautiful, perfect healthy boy!

During the cycles I did acupuncture, which I LOVED!! I also believe the women's health is a big factor. Eating healthy, maintaining a healthy weight, exercise and being emotionally prepared.

This process was one of the biggest areas in my life that I was able to grow as a person. It causes you to check who you are; your morals, emotions, relationship and can either make you stronger and wiser or (if you allow it to) can cause you to make decisions based on emotions instead of principles. And not just with the clinical process, but in your marriage, with your friends, with your finances, and many more!

We are not planning on doing any more treatments although I would love to naturally get pregnant with one more child.

The advice I would give to someone starting this process is I would hope that they believe in the Giver of life! Because the process can tear you up and you must have a foundation of belief that God has a plan for you and your family, whether the process works the first time, second time or not at all!

Kara

We tried for 12 months to get pregnant. We already had another child, conceived naturally, and it took us about 6 months to get pregnant with that child. I was 37 years old. After 12 months with no success I was referred to a local IVF clinic by my OB/GYN who felt I needed further tests.

At the IVF clinic we did a multitude of tests, one was to find out if my fallopian tubes were clear and some blood tests too. It seemed like there was a lot of testing done. We also tried artificial insemination (IUI). We decided to do IVF because of my age. We thought it was the right thing to do. I was convinced that it was going to be the only way I could get pregnant.

I was lucky enough to only go through one IVF cycle. However, the cycle went on longer than the typical cycle because I had to continue to give myself shots since my body wasn't producing many eggs. They got 6 eggs, and 3 became viable embryos. We transferred all 3. One implanted and I got pregnant!!! I had a beautiful baby boy. He is now 5 years old.

The hardest part of doing the treatments was the emotional toll it took on me. I was very stressed and anxious during the process. It was very expensive, so I was concerned about that as well. Our insurance did not cover any part of IVF. Everyone at the clinic was great and very thoughtful, but I was so hopeful for a pregnancy that I worried myself sick at times. The thing that surprised me the most was how strong I actually was, that I made it through the process. The thought of giving myself shots, and my husband giving me shots, really frightened me. I found out that I could handle it just fine.

I want to tell people doing fertility treatments to try and relax sometimes. Don't let it consume you. There is life outside of trying to get pregnant. Do fun things with your husband to get your mind off of things. It really takes a toll on your sex life because it becomes a chore, not something fun.

Allison

I tried to get pregnant for 7 months when I was 30 years old without any success. I went to see my OG/GYN who ran some tests and when my husband was found to have a low sperm count my doctor referred us to an IVF clinic. At that stage we were told our only option was IVF with ICSI. I did two cycles of IVF. During the first cycle I got 9 eggs, 6 were mature and 4 fertilized with ICSI. I did a day 2 transfer of two good quality embryos; the other two embryos we had weren't good. I got pregnant with a singleton but had a miscarriage just before 9 weeks.

On the second IVF cycle I got 6 eggs, 6 were mature and 4 fertilized. Again we did a day two transfer of 2 good quality embryos. This time I got pregnant with fraternal twins. One of the babies started to show growth problems at 18 weeks along and passed away 3 weeks later (never grew past a 15 week size). Twin A is doing well and I am almost 24 weeks pregnant.

During the IVF cycle I tried to not get excited during the stimulation phase—you go through so much with the shots and monitoring appointments that it is too hard to deal with if you don't have a good outcome. Also, there is the loss of control you have over the cycle–with the constant worry that each time the nurse phones you they might have bad news. Then the two-week wait is hard because it is so long. There is nothing more that you can do and you just pray and hope each second that it worked and that your embryos are still growing and burrowing in you after the transfer, but you don't know and won't know until enough time has passed. I also didn't have much support from my family during the whole process (my mother is a strict catholic and is against us doing IVF) and that was very hard not being able to talk to her and having her judge us.

I was, however, surprised how the process wasn't as bad as I thought it would it be. The day I dreaded the most was the egg retrieval and it was actually the easiest day of them all. I even looked forward to that day with my second cycle because of the nice anesthetic and pain meds!

One harsh reality I have come to learn is that a positive pregnancy test does not equal a baby. All I wanted for a year was to get pregnant and when I did after my first IVF I was so happy and thought the worst was over—thinking pregnancy loss isn't as likely as not getting pregnant. After experiencing a miscarriage I knew that getting a positive test with nice increasing betas and a heartbeat on an ultrasound means nothing and that you are not out of the woods until you have the baby in your arms.

To people reading this; know that you are not alone. There are many other women out there going through the same thing as you. I felt embarrassed when we started because I didn't want people to think that there was something wrong with me. I soon realized that there are thousands of women doing IVF for all kinds of reasons and there is no reason to feel ashamed.

Kelly

After two years of trying to get pregnant we went to see my OB/GYN who referred us to an IVF clinic. I was 38 and my husband was 33. We went to the first clinic on the recommendation of my OB/GYN and a second clinic after the treatment failed. We decided we needed to do our homework in order to be sure we were actually at the best clinic, not just going there based on a recommendation with no homework.

The initial testing they did was pretty standard; HSG, FSH, semen analysis, thyroid testing, nothing too invasive since I had children from previous marriage. The doctors assumed my problem was due to a luteal phase defect and that I needed to be on clomid. We only did IUI's for the first 2 years. Eventually the RE said we needed to move to donor egg, as it must be my old eggs. Instead of doing our homework we believed him and wrapped our minds around donor egg. During the first donor cycle we got 13 eggs, quality was ok; we did a day 3-day transfer with fresh embryos and a day 3 transfer with frozen embryos. The pregnancy tests were negative.

For our second cycle I read and read and found a few things that I wanted to test before cycling again. The doctor refused to do them at first, he said we didn't need to do them as studies show that women with kids typically have and keep the beta-3 integrins and so he thought the test would be normal, but he did respect my need to rule it out. It came back abnormal.

We ultimately did another IVF cycle with donor eggs and with the fresh transfer we put back 3 embryos. We got another negative result. For a frozen embryo transfer we put back 4 embryos and hit the jackpot, I am currently 19 weeks pregnant with twins.

When we found out we needed to use donor eggs we had to decide what is important—having a child or having a biological child. If the goal is to be a parent, then donor egg is a great option and was very easy for me to embrace. We could have adopted, and in a way donor egg is a form of adoption; we just got to start the process at conception rather than birth and we were able to at least have the DNA connection to my husband.

Choosing a donor for me was easy—I wasn't that picky. Set your priorities, but at the same time remember no family is perfect as much as we want that from our donor. For our first cycle we scoured the agencies for the perfect donor—when it didn't work we realized that perfect wasn't necessarily going to get us our child and therefore how perfect was she? For cycle number two we chose a clinic with great stats, and an in house donor pool and chose among their donors. Our choices were obviously more limited, but we were able to find a donor that was right for us, and in the end we did get pregnant and now expecting two blessings in just a few months. To us, that makes her perfect.

When starting the process, determine your budget, search the SART and CDC sites, and choose the clinic with the best rates that you can afford. If you can do a success guarantee program, take advantage of it. Remember, it is just as easy for you to travel to a clinic as it is for a donor to travel to you, but traveling to the clinic is MUCH less stressful, especially if going to one of the top rated clinics. You have lots of options, find the forums/boards out there like IVFC, PVED, etc.. Join, read, research, and use the experience of others. It is truly invaluable and can save you lots of time, money and heartache by using that knowledge.

The hardest part was putting our trust and faith in a doctor and believing he cared about us and I was more than just a paying customer. What surprised me the most was that the IVF doctors wanted us to just try again after our failures. No testing, just the excuse of bad eggs from the first donor. And then, when we decided to talk to other clinics they were rude and offered no help—even getting my records was like pulling teeth. They wouldn't help with out of town monitoring when we did decide to switch clinics. It was as though we had offended them.

I wish I knew from the start that I could travel to any clinic I wanted to, rather than thinking I needed to be at a clinic close to home and the donor needed to come to me

We still have eight frozen embryos and if this pregnancy continues well and we are OK with raising the twins we would consider possibly doing another cycle in the future. We won't be doing any more fresh transfers.

My advice is to decide what's important, get advice and recommendations from as many people as you can, don't waste time, and take matters into your hands. Research immune issues, look at possible tests, and make your RE listen to you. If I hadn't done that for our second cycle I don't think I would be pregnant now. I think I would still be scratching my head wondering why it doesn't work. I would have moved on to surrogacy, which would have been ok, but being able to carry these babies has been so amazing that I can't imagine now having missed out on it.

Marcia

I was never on birth control pills and started trying to get pregnant when I was 29 years old. I started my infertility work-up with a doctor at age 30 and had a hysteroscopy done. I eventually started doing IUI's when I was 32 years old.

I did a lot of testing including FSH blood levels and a hysteroscopy ....by the time I was 37 we decided not to do a laparoscopy to see if I had endometriosis but instead went straight to IVF.

I did one cycle of IVF at 37 years old in 1996. They harvested 11 eggs, 8 fertilized, 4 were transferred. We have a Polaroid photograph of the embryos...2 eight cell, 1 six cell and 1 four cell. We decided to put back all of them because of my age. The outcome was a singleton pregnancy, which delivered at term with a healthy girl, 7 lb. 8 oz.

I think the hardest part about doing IVF was waiting for the CVS test to come back telling us whether the fetus was healthy or not. I didn't mind any of the shots, meds, procedure…I was ready to try again if I needed to. I did have a very heavy, clot-filled bleed about four weeks after the transfer right before the first ultrasound, which made us think that I miscarried. I went in afterwards for the ultrasound and we saw the fetal heartbeat! That was a wonderful sight! Since these types of bleeding episodes happen with IVF patients, we think it may have been another embryo that did not implant which I passed.

I wish I knew how much I enjoy being a mom and that I should have tried again to do IVF before I turned 42 or 43. My daughter really wanted a sibling too.

I would like to tell people don't give up…there is always a way to have a child, whether it's with your own eggs, donor eggs, donor sperm, surrogate, or adoption if all else fails. If not becoming a parent will be the one regret in your life, then you must try, since the technology, the scientists, the specialists and the acceptance is all there and better than ever.

Sarah

I waited too many years trying to get pregnant naturally, we held off talking to a doctor due to work restraints and the cost of fertility treatments. I was 32 years old when we eventually took the plunge. We had some friends who were using the clinic we chose and the doctors were on our insurance preferred provider lists so insurance would cover a portion of the treatments.

We started testing with Kaiser and found we had a lower sperm count and blocked tubes. There were quite a few blood tests in there as well. I wish I had had a better sense of what to look for and expect at the time as I would have recorded and tracked the tests better.

With my own eggs we did an IVF cycle and we had 11 eggs retrieved, 9 mature and 9 fertilized, however, by the time of the 5 day transfer only one remained. In later discussions with our doctor even that one was not of day 5 quality. We transferred that one and I had a negative pregnancy test. I chose to do acupuncture with the cycle, even though my tests showed I would be ok if I did not do this. I had decided I would do everything possible to give us the best chance at success and loved it!

Now we are getting ready to do our first (fingers crossed only) donor egg cycle. We both knew that children were what we wanted though we could have gone the adoption route, which is still not out of the question if needed. We wanted to have the bonding experience with the baby as well as knowing how the baby was being cared for before its birth. With adoption there is no way to be sure the birth mother is caring for herself the way she should or, even worse, change her mind after we had bonded with the baby. Using donor egg was not at all a difficult choice. The decision was made in our minds before we left the clinic after the consult regarding the failed cycle; the question was primarily how to pay for it!

At the IVF center we were provided with the clinic donor site prior to leaving the consult. When we got home I started to look through the files. One donor stood out to me, I selected a few with similar eye colors and hair and showed them to my husband. My husband pointed to her and said "I like her best because your smiles are very similar and I have always liked your smile."

My advice to people starting a donor egg cycle is to do what is best in your heart! Don't let others opinions influence your decision and don't be afraid to talk with someone if you need to.

For me, the biggest fear about doing IVF was every time I would go in for blood work and/or ultrasound, I would be told it might never happen and I need to stop everything and look into other options.

The thing that surprised me the most is that I can actually tolerate the shots! I can't give them to myself but I'm not terrified anymore! If I had known earlier that my insurance covered both medical treatment and prescriptions I would have started this process years ago!

My advice is to do your research through all of the web sites such as SART and Resolve and be ready with questions. Don't be afraid to ask questions of the doctors, nurses or anyone else at the clinic; if you don't feel you can ask a question then ask yourself "is this the right place for you?"

Mary

My husband and I tried for eight years to conceive before doing IVF. We tried several cycles of IUI first. Since my husband has a daughter we figured it would just take time. No luck. My doctor finally referred me to an IVF clinic; she actually knew the RE personally and they took my insurance.

I have unexplained infertility, but using ultrasound we came to the conclusion that my uterine lining wasn't thick enough for an egg to attach to and stick. After just four treatments of acupuncture the doctor noticed incredible changes in my uterine lining. That was amazing to me, I was so glad to hear it.

After that we went on to do an IVF cycle and retrieved 14 eggs. We ended up with 7 normally fertilized. We decided to put two back in, they both held on and now we have boy/girl twins born in 2008. They are totally healthy and I had the best pregnancy with no real concerns. I carried the pregnancy to full term and was induced at 38 weeks.

I was surprised how easily everything went with the procedure.. It was smooth and comfortable. We started the whole process in September 07 and by November I was pregnant. It was truly a blessing. I procrastinated about it for five years and when we finally decided it was time I was 34.

I have two stepdaughters, and the twins so we will not be having any more babies! My husband had a vasectomy just in case there was a chance that I may get pregnant on my own!

My advice for anyone else... The time has to be right for each person. It was a great experience for my family and me and I know everyone doesn't get the same results. Follow your heart and just know that whatever happens, you did all you could and there will be no questions of what might have been. That was always how I felt. I knew choosing IVF was a good decision for us and I needed to at least try. I thank God everyday; these babies have brought my blended family so close and that is the best gift of all!

Amy

At age 30 my husband and I decided to start trying to conceive. After one year with no success my friend recommended a doctor she knew. She had gone for an initial consult and then got pregnant on her own. The clinic was on my insurance and close to my office (in NYC), so it made sense, as a first step, in a lot of ways. I have remained with the same clinic since that day.

The doctor ran basic fertility tests on husband and me. We found no reason for why I wasn't getting pregnant and the doctor thought that based on my age I might be pregnant by the end of the year. We did 6 IUI's (with clomid and Gonal-F) with no success, before moving onto IVF. After that many unsuccessful IUI's, coupled with my (relatively) younger age and unexplained diagnosis, my RE felt that we would get a better explanation of why it wasn't working with IVF.

Here is a summary of my treatments:

IVF #1: I was part of a study group testing a new single dose of stimulation medication. We did this because I was eligible for it ("young", new to IVF and unexplained) and my insurance did not cover IVF, so it cost a lot less to participate in this at my doctor's office. The quality of embryos wasn't great (two transferred), and the result was negative. There was 1 blastocyst frozen from this cycle.

IVF #2: The doctor wanted to cycle me with regular meds, and I got new insurance, which would cover the cost (hurray!) Again, the embryos did not look good (three transferred this time), and I had a very low beta number that dropped. There were no embryos left for freezing. After this cycle, the clinic performed a karyotype blood test. It revealed a chromosomal balanced translocation and MTHFR/thrombosis mutation. We were relieved to receive a diagnosis; however, we didn't realize that it was such a difficult one to overcome. After meeting with a genetic counselor, she persuaded us that cycling again with PGD would help find any healthy embryos but that some people may not have any at all.

IVF #3: This was the last cycle that insurance would cover. A probe was made for the chromosomal translocation to do the PGD test, but when the time came, we were told that the embryos were not good enough quality for PGD. So they transferred the best 2 as well as the one that was frozen from round #1. Negative. Based on the scrambled looking pictures of embryos we received each transfer day, and the latest results, I started looking into donor eggs as a solution. We went to counseling regarding donor eggs. The counselor advised us that my older sister is the person closest to my DNA in the world, so my husband, who had initially been reluctant to the idea, agreed to cycle with my sister.

IVF #4 Donor eggs: We cycled with my 37-year-old sister (I was 33 at the time)—she has three healthy kids of her own, with no fertility issues. We transferred three good quality embryos. None were left for freezing. I was initially pregnant with twins, however, one stopped developing early on. Our beautiful baby girl was born in July '08. In '09 I stopped breastfeeding my daughter at nine months old for my periods to return to try to cycle with my sister again for a sibling. We were very hopeful, since we had success with my sister on our first attempt cycling together.

IVF #5 Donor eggs: We cycled with my sister again, now 39 years old, and transferred three good quality embryos which resulted in a low beta number that dropped to zero. Two embryos were frozen.

FET #6): We transferred the two frozen embryos, which resulted in a low beta number that rose slightly, reached a plateau, and dropped. Negative. We were extremely disappointed that our attempts to have a second child failed. We are currently discussing whether or not to cycle with my sister again. My husband is not interested in growing our family through an anonymous donor, adoption, or any other method. My doctor agreed that we could cycle with her again until her 40th birthday. He will try Lovenox to deal with my MTHFR/thrombosis mutation, instead of just baby aspirin and Folgard, which we used previously. The doctor believes the cause of the low dropping betas could have stemmed from the egg quality as well.

Typically we transferred two or three (usually three) embryos on day three, based on the doctor & embryologist's recommendations. If we cycle again, the doctor indicated that he would not save any embryos for freezing, but would transfer all viable embryos on transfer day (with my sister's eggs, that has typically been 7–9) and see if any "stick".

We used my sister as an egg donor because we felt like we weren't making progress with my own eggs, and we wanted more of a sure thing, since it would be out-of-pocket. The decision was more difficult for my husband to make than me initially. He wanted the baby to be a part of me, and resolved to take the "next best thing." I wanted a baby, and didn't care whether it was genetically linked to me or not. He would not and will not consider an anonymous donor, although we have discussed it. I don't know what advice to offer, other than you'll know it's the right time to take that path when there aren't solutions down the paths that you've tried already. We seemed to hit a place emotionally and with the logic of understanding where our dollars were being spent that helped to dictate our decisions.

The hardest part is not knowing when it will end. Not knowing if or how it will end. Receiving a progesterone shot that made it hard to walk for several weeks was difficult, but part of the necessary pain. And I hate to say it, but the tens of thousands of dollars that we have worked hard for, with nothing in return, are difficult to lose. Wanting something that 9/10th of the population has no problem "acquiring", and putting all of your heart, soul, emotions, time, and money into making it happen, with an unknown (and generally negative) outcome is hard. There are few things in life that you can devote yourself to so completely and still not gain any reward, and IVF is one of them. I worked hard to do my best at a top prep school, an Ivy League college, and two graduate degree programs. No matter how hard I worked at trying to have a baby, it didn't matter. I would be willing to put myself through any number of shots, pills, and 5:30am wake ups for monitoring, and a depletion of our savings, if I only knew "On [date] you will become pregnant with a viable pregnancy." I tell myself that it gave me more patience for parenting. I didn't realize that these feelings would resurface so strongly when attempting it again for a second child. I have some solace in the joy of having my daughter in my life. I also have many unresolved feelings knowing that my husband and I would both like to have two children, but also think that maybe we should work harder to be happy with just one.

My biggest surprise was other people's insensitivity to what I was going through when I chose to share my experiences with them. We shared our experiences with a few close family and friends, and several of them didn't understand why we wanted to limit the spread of the information. I don't appreciate being the subject of gossip, and that is how I felt, when my mom, for example, would feel the need to tell her friends about my IVF journey. After I became pregnant, another friend dismissed my infertility and said that it didn't matter anymore, that I was pregnant now, like she had been after one month of trying on their own. It wasn't the same to me though; being pregnant after trying for 1 month versus trying for almost 3 years—countless medical procedures and dollars later. It is difficult to explain this to someone who hasn't gone through it. The experiences that I have gone through have changed who I am. I still feel the pain of infertility, even though I have a beautiful daughter in my life now.

If I had a crystal ball, I really wish I would have know that I would have a beautiful, healthy baby girl in July 2008, as well as a date for #2, if there is to be one.

Medically, I know this isn't a realistic answer. If I had walked into the doctor's office and we had gone straight to donor eggs, I would have a lot more money in my pocket and a baby sooner. But, I know that it didn't make sense from his perspective and we certainly would have been taken aback to do this immediately. I wish they had done the karyotype blood test sooner though. Being "1 in 625" with this condition, and having a severe form of it, I know that they didn't have a reason to do it until they started seeing chemical pregnancies in my medical records.

My advice to someone reading this is to stick with it. It will happen. You don't know when or how, but the family you always wanted will be yours, if you keep doing everything in your power to make it happen.

Vivienne

This story begins after I had a miscarriage. In the recovery time, I had a hysteroscopy and D & C for uterine polyps. I also re-engaged acupuncture and found a wonderful lady who specialized in infertility and women's health. She was a Godsend!!! She helped me to heal both spiritually and physically and got me back on the good path.

During that time, we also refinanced our home and took out a second mortgage for fertility purposes. We decided that we wanted to go no holds barred, all guts or no glory, and give IVF a try. We wanted our baby and we wanted him now. The force of the universe proved to be with us, and we landed donated medications through our clinic! Four months after our miscarriage, we started our first IVF cycle.

I don't know from where my dogged optimism grows. Despite all the setbacks I'd had, I still believed that we'd breeze through IVF: develop between 12–15 beautiful follicles, the vast majority of which would go on to become thriving embryos; have our pick of blastocysts for a 5-day transfer; freeze the rest of the brood for siblings. The worst, I thought, would be that I'd develop OHSS from growing so many eggs. Can you guess where this is going?

I had 9 resting follicles at my base count. "Nine," I asked, "is that good?" My question was gracefully deferred. That set the pace for my cycle, as every intimate appointment with the "love wand" brought me face to face with my depleted reserves. I was 33 and apparently had the ovaries of someone much older, or at least that was how I was responding. Who knows why this happens, I was told, maybe genetics. Genetics. My mom was the oldest of 8 siblings who were born for breeding. I have one cousin who is 15 months older than me who had 7 children by the time she was 32. But I digress.

My cycle was facing cancellation. The cutoff was five follicles, and I was on the fence. Since we were 100% self-pay at this point, wouldn't it be more prudent to pull out now, convert to IUI, hedge our bets and try a different protocol next time, one for low responders? I went to acupuncture after that noteworthy appointment with this decision heavy on my heart. I knew I'd disappointed my husband, who was trying his best to be strong for me. But even he had to collect himself in the car before he could see to drive home. Acupuncture was supposed to be relaxing, but when I let my guard down for one second the dam burst. I heaved noisy and pathetic sobs, spewed forth mountains of mucus, and despite my practitioner's assurances I snuck out the back door at the end of the session, with a promise to pay next time. I'm still both grateful and embarrassed to this day.

I thought about it. My husband thought about it. We talked about it. I asked my friends on RESOLVE their opinions. And in the end everyone came to the same conclusion: go forward. Don't stop now. So I didn't.

I had the HSG shot on November 2, which was a nationwide voting day. I considered that with this shot I'd cast my vote on hope, on optimism, on overcoming adversity, on faith, and on the notion that I, too, was worthy of a happy ending. "Please," I begged, "just let me make it to embryo transfer. Give me a chance."

Egg retrieval day was on a Thursday. I was pink-cheeked and so chuffed to be there. I gulped at the anesthesia and asked God to help me. I knew that the first thing I'd hear upon waking would be how many eggs they had retrieved. Before I blacked out, I tried to steal myself for the possibility of one. I knew that some follicles prove empty, especially when you've been stimming as long as I had. Two would give me twice the chances. Every egg would count and I'd be grateful. Three was my lucky number.

"Six," I heard the nurse whisper, before I even knew I was awake.

I floated home with the knowledge that my half dozen half babies were in wonderful hands in the embryology lab. I felt so accomplished, considering my parameters. It was out of my hands. I did the best I could.

The fertility report delivered the next day froze the blood in my veins. Two had fertilized. Emergency ICSI was not a good option. I was to come in the next morning for a 2-day transfer, pending the survival of my duo over the next 24 hours.

At this point, a certain grace started to come over me. I couldn't cry anymore; it just didn't feel appropriate. Something told me to be strong. I will admit, I did hold my breath the following morning, expecting the call to tell me it was off, that none had survived. The call never came.

We had a day 2 transfer of my 2 embryos on a Saturday morning in the first week of November. The weather was crisp and sentimental. My acupuncturist met us there, to perform a session before and after transfer. The embryologist broadcast an image of my 2 embryos before they transferred them into me. My babies, I thought, stay with me. My acupuncturist told me that I looked different, when they wheeled me out. I felt different. I felt so grateful.

I'll skip the harried details of my two week wait: the bed rest that was actually physically challenging (I wasn't used to being so sedentary), the impromptu toilet my husband lovingly rigged up for me in the basement so I could watch the big screen, the wondering, doubting, hoping... In the end, I skipped the home pregnancy test and went straight to the doctor's office for beta HCG on day 12p ET. The night before, my body had already told me I was pregnant. Our entire clinic called us on speakerphone, with my husband taking the call. I heard him say, "I'm going to vomit," and I knew. I knew. I started sobbing again, but this time with pure joy. And that was the end of our infertility journey, and the beginning of our next adventure, to become a family.

Claire

I was 31 when we started trying to get pregnant, in 2006. I was charting my temperature hoping to maximize our chances of getting pregnant sooner. After six months, it seemed like I was ovulating and we were timing intercourse well, but I wasn't pregnant. I saw my OB/GYN, who advised me to stop charting (i.e. just relax! Argh!) and come back if we still weren't pregnant in a year. It occurred to me that we had lots of data on my cycles, but knew nothing about my husband's fertility. He saw his PCP and asked for a semen analysis, and we had our answer. We were referred to a local IVF clinic by my OB/GYN. I looked online at clinic statistics and infertility message boards, and felt that this was the best choice that we could make.

At the clinic they ran repeated semen analyses, as well as the usual battery of tests on me (HSG, FSH, etc), all of which were normal. Because of my husband's very low sperm count, low motility, and poor morphology, we were advised to proceed directly to IVF. Fourteen eggs were retrieved. Thirteen were mature, nine fertilized, and six made it to blastocyst. We transferred two embryos and I became pregnant with our daughter, now 2 1/2. The remaining four blastocysts were frozen.

For our FET cycle this summer, we transferred one embryo, after thawing two of our embryos, because we did not want to become pregnant with twins. I became pregnant, but with a blighted ovum. So, for our upcoming IVF, I am again thinking of transferring two, depending on how the thaw goes.

I did acupuncture, during all my treatment cycles. I don't know if it improved our chances of pregnancy but it did help me to de-stress.

The hardest part of fertility treatments for me was the significant emotional and financial investment that goes into this one chance to get pregnant. And knowing that if it doesn't work, there's only more emotional and financial investment to come.

What surprised me the most was that the shots really weren't that hard to do. I have always been a wimp when it comes to needles (so ironically, getting pregnant and having a baby involved, for me, acupuncture, injectables, and then blood sugar monitoring during pregnancy—lots of needles!), but I wanted to get pregnant so badly that jabbing myself with a needle barely registered with me, once I got over my initial fear.

Looking back, I see how incredibly lucky we were that we were diagnosed so quickly, able to move on to IVF so quickly, and got pregnant on our first try. Amazingly, we were pregnant within a year of first trying to conceive.

My advice to people starting on this journey is to remember to take care of yourself during the cycle: massage, housekeeper, long walks, long baths, etc—whatever minimizes stress and promotes relaxation—not to increase chances of pregnancy, but to keep yourself sane. It's a stressful process emotionally, physically and financially. Find support, either from friends (preferably someone who has also been through IVF) or online support groups and be as informed as you can be.

Resources

American College of Obstetricians and Gynecologists (ACOG) Resource Center
(800) 762–2264
www.acog.org

American Fertility Association
(888) 917–3777
www.theafa.org

American Society for Reproductive Medicine
(205) 978–5000
www.asrm.org

Fertile Hope
(888) 994-HOPE
www.fertilehope.org

Food and Drug Administration (FDA)
1–800–994–9662
www.fda.gov

Genesis Genetics
313–579–9650
www.genesisgenetics.org

International Premature Ovarian Failure Association
(703) 913 4787
www.pofsupport.org

International Council on Infertility Information Dissemination, Inc.
(703) 379–9178
www.inciid.org

IVF Connections
www.ivfconnections.com

National Embryo Donation Center
(866)-585–8549
www.embryodonation.org

Parents Via Egg Donation (PVED)
(503)-987–1433
www.parentsviaeggdonation.org

Reprogenetics
(973) 436–5015
www.reprogenetics.com

RESOLVE: The National Infertility Association
(888) 623–0744
www.resolve.org

Snowflake
(714) 693–5437
www.nightlight.org/adoption-services/snowflakes-embryo/

Society for Assisted Reproductive Technology (SART)
(205) 978–5000
www.sart.org

The Centers for Disease Control Division of Reproductive Health
(800) CDC-INFO
www.cdc.gov/Reproductivehealth/DRH/index.htm

The Endometriosis Association
(414) 355–2200
www.endometriosisassn.org

The Polycystic Ovarian Syndrome Association
www.pcosupport.org

Glossary of Terms

A

Adhesions

Scar tissue in and around the inside of the pelvic region. Adhesions may interfere with transport of the egg and / or implantation of the embryo in the uterus.

Amenorrhea

Refers to a woman who does not have a menstrual period.

Anovulation

The failure to ovulate

Antibodies

Proteins made by the body to attack or fight foreign substances. Antibodies normally prevent infection, however, they can be made against sperm, sometimes causing fertility problems. Either the male or female partner may produce sperm antibodies.

Antisperm antibodies

Antisperm antibodies attach themselves to the sperm and can inhibit movement. In men, this is sometimes in response to injury or surgery to the testes when the blood-sperm barrier has been breached. Antisperm antibodies may impair the sperms' ability to fertilize an egg.

Assisted Hatching

A small hole is made in the outer shell around the embryo (the zona pellucida) before transfer to the uterus.

ART (Assisted Reproductive Technology)

A term used to describe advanced scientific interventions, such as IVF, which are used to treat infertility.

Artificial Insemination (Intra-Uterine Insemination, IUI)

A type of fertility treatment in which sperm is washed and injected directly into the uterus around the time of ovulation

B

Blastocyst

A stage of embryo development approximately five or six days after egg collection

Blastomere
A single cell within a pre-implantation embryo up to the morula stage of development. Each blastomere is totipotent and has the ability to develop into a whole embryo if it was removed and grown on its own

Blighted ovum
A non-viable pregnancy that shows a gestational sac with no fetus on ultrasound

Beta hCG test
A blood test used to detect early pregnancy and to monitor progress of the pregnancy during the first weeks.

C

Canceled cycle
An ART cycle in which ovarian stimulation was carried out but was stopped before eggs were retrieved. Cycles are canceled for many reasons: there may be too few or too many eggs developing, the uterine lining may not be optimal for embryo replacement, the patient may become ill, or the patient may choose to stop treatment.

Chromosome
A thread-like strand of DNA in the cell that carries genetic information. Humans have 23 pairs of chromosomes in each cell of their bodies. Mature eggs and sperm have 23 single chromosomes.

Clomiphene Citrate (Clomid)
A fertility drug taken orally that stimulates ovulation through the release of gonadotropins from the pituitary gland.

Compaction
A stage of early embryo development between the 8-cell and the blastocyst stage, around day four of in vitro culture. Describes when the cells merge together to create a morula.

Corpus Luteum
The follicle from which an egg is released. The Corpus Luteum produces progesterone, which is responsible for preparing and supporting the uterine lining for implantation.

Cryopreservation
Freezing of biological material such as eggs, embryos and sperm.

Cumulus cells
A protective layer of cells surrounding the egg

D

D & C (Dilation & Curettage)
A minor surgical procedure in which the physician first dilates or opens the cervix and then inserts a spoon shaped instrument to remove material from the internal lining of the uterus. Dilation stands for opening the cervix and curettage means the scraping of the uterine wall.

Donor egg cycle
An embryo is formed from the egg of one woman (the donor) and then transferred to another woman who is unable to use her own eggs (the recipient). The donor relinquishes all parental rights to any resulting offspring.

Dysmenorrhea
Painful menstruation; may be a sign of endometriosis.

E

Ectopic pregnancy
A pregnancy in which the fertilized egg implants in a location outside of the uterus; usually in the fallopian tube, ovary, or abdominal cavity. Ectopic pregnancy is a dangerous condition that must receive prompt treatment.

Egg retrieval (oocyte retrieval)
A surgical procedure to collect the eggs contained in the ovarian follicles

Embryo
An egg that has been fertilized by a sperm and undergone one or more divisions.

Embryologist
A highly trained scientist specializing in reproductive laboratory procedures

Embryo transfer
Placement of embryos into a woman's uterus through the cervix after in vitro fertilization

Endocrinology
The study of the body's hormone secreting glands

Endometrium
The lining of the uterus that grows thick each month and is lost through menstruation or remains intact to nurture an embryo if conception takes place

Endometriosis
A medical condition, which may contribute to infertility, involving the presence of tissue similar to the uterine lining in locations outside of the uterus, such as the ovaries, fallopian tubes and abdominal cavity

Estrogen
Female sex hormone produced by the ovary.

F

Fallopian Tubes
Ducts through which eggs travel to the uterus from the ovary after they are released from a follicle. This is where fertilization normally occurs inside the body when sperm swims up the tube and meets the egg.

Fibroid
A fibrous non-malignant tumor in the uterus. May affect fertility.

Fertilization
The penetration of the egg by the sperm and the resulting combination of genetic material before an embryo is formed

Fetus
The unborn offspring from the eighth week after conception to the moment of birth

Follicle
A fluid-filled structure in the ovaries that contains a developing egg

Follicle Stimulating Hormone (FSH)
A hormone produced by the pituitary gland that stimulates egg maturation in the ovaries.

Fragile X
A genetic disorder caused by the mutation of a gene on the X chromosome. Women may have premature ovarian failure, high FSH or a poor response to gonadotrophin stimulation.

Fragmentation (of embryos)
When small fragments of cell are lost during cell division in the pre-implantation embryo. Microscopically, this gives the embryo a poorer appearance, leading to a lower grade. Fragmented embryos are less likely to implant than those without any fragmentation. We will always choose the least fragmented embryos for transfer.

Frozen embryo cycle
An ART cycle in which frozen (cryopreserved) embryos are thawed and transferred to the woman

G

Gamete
A reproductive cell. Either a sperm or an egg

Gestation
The period of time from conception to birth

Gestational carrier (a gestational surrogate)
A woman who carries an embryo that was formed from the egg of another woman. The gestational carrier usually has a contractual obligation to return the infant to its intended parents.

Gestational sac
A fluid-filled structure that develops within the uterus early in pregnancy. In a normal pregnancy, a gestational sac contains a developing fetus.

Gonadotropins
Hormones that control reproductive function. Human menopausal Gonadotropin (HMG), Follicle stimulating hormone (FSH) and Luteinizing hormone (LH).

Gonadotropin Releasing Hormone (GnRH)
A substance released from the hypothalamus in a pulsatile manner approximately every 90 minutes. This hormone acts on the pituitary gland enabling it to secrete LH and FSH, which stimulate the gonads.

H

Hatching (of embryos)
The process of an embryo escaping from its outer shell (zona pellucida) prior to implantation in the uterus

Hirsutism
Overabundance of hair growth often found in woman with excess androgens

Human chorionic gonadotropin (hCG)
A hormone produced during early pregnancy that keeps the corpus luteum producing progesterone. hCG is also used to trigger ovulation during fertility treatments.

Hyperthyroidism
Overproduction of thyroid hormone by the thyroid gland. This leads to an increase in metabolism and can cause estrogen to 'burn up' too rapidly, thereby interfering with ovulation.

Hypothalamus
A part of the brain that is the "hormonal regulation center." The hypothalamus is located adjacent to and just above the pituitary gland.

Hypothyroidism
Underproduction of thyroid hormone by the thyroid gland. The resulting lowered metabolism can interfere with the normal breakdown of hormones and may lead to lethargy. Women may suffer from elevated levels of prolactin and estrogen, which can interfere with fertility.

Hysterectomy
Surgical removal of the uterus

Hysterosalpingogram
A procedure used to assess the anatomy of the cavity of the uterus and the fallopian tubes.

Hysteroscopy
A thin telescope passed through the cervix to visualize the inside of the uterus.

I

ICSI (intracytoplasmic sperm injection)
A procedure in which a single sperm is injected directly into an egg by an embryologist. This procedure is most commonly used to overcome male infertility problems.

IUI (intrauterine insemination)
A type of fertility treatment in which sperm is washed and injected directly into the uterus around the time of ovulation

IVF (in vitro fertilization)
An ART procedure that involves removing eggs from a woman's ovaries and fertilizing them outside her body. The resulting embryos are then transferred into the woman's uterus through the cervix.

K

Karyotyping
A test performed to analyze chromosomes for the presence of genetic defects.

Klinefelter's Syndrome
A genetic abnormality characterized by having one Y (male) chromosome and two X (female) chromosomes. This condition may cause a fertility problem.

L

Laparoscopy
A surgical procedure in which a fiber optic instrument (a laparoscope) is inserted through a small incision in the abdomen to view the inside of the pelvis

Luteal Phase
Post-ovulatory phase of a woman's cycle. The corpus luteum produces progesterone, which causes the uterine lining to thicken to support the implantation and growth of the embryo.

Luteal Phase Defect (or deficiency)
A condition that occurs when the uterine lining does not develop adequately due to inadequate progesterone stimulation or because of the inability of the uterine lining to respond to progesterone stimulation. LPD may prevent embryonic implantation or cause an early miscarriage.

Luteinizing hormone (LH)
The mid-cycle release of LH that causes an egg to be ovulated. Ovulation detection kits detect the sudden increase in LH signaling that ovulation is about to occur, usually within 24–36 hours.

M

Male factor
Any cause of infertility due to low sperm count or problems with sperm function that make it difficult for a sperm to fertilize an egg under normal conditions.

Miscarriage (also called spontaneous abortion)
A pregnancy ending in the spontaneous loss of the embryo or fetus before 20 weeks of gestation

Morula
A stage of embryo development after four days of culture between the 8-cell stage and the blastocyst stage of development when all the cells of the embryo merge together

Motility (of sperm)
The ability of sperm to move and swim normally

Multi-fetal pregnancy reduction
A procedure used to decrease the number of fetuses a woman carries and improve the chances that the remaining fetuses will develop into healthy infants. Multi-fetal reductions that occur naturally are referred to as spontaneous reductions.

Multiple-infant birth
A pregnancy that results in the birth of more than one infant

Multiple-fetus pregnancy
A pregnancy with two or more fetuses

O

Oocyte
The female reproductive cell, also called an egg.

Ovarian Cyst
A persistent fluid-filled sac inside the ovary. Cysts can produce hormones that interfere with ART cycles.

Ovarian Hyper Stimulation Syndrome (OHSS)
A potentially life-threatening condition following ovulation induction treatment. OHSS arises when too many follicles develop and hCG is given to release the eggs. This condition may be prevented by withholding hCG when ultrasound monitoring indicates that there are a large number of follicles in the ovaries. IF OHSS is evident during an IVF cycle, all the embryos may be frozen at the pro nuclear stage (the day after the egg collection) and replaced at a later date when the condition has subsided as pregnancy can exacerbate the symptoms of OHSS.

Ovarian monitoring
The use of ultrasound and/or blood or urine tests to monitor follicle development and hormone production.

Ovarian stimulation
The use of drugs (oral or injected) to stimulate the ovaries to develop follicles and eggs

Ovulation
The release of an egg from an ovarian follicle

Ovulatory dysfunction
A cause of infertility due to problems with egg production by the ovaries

Ovum
Another name for the egg

P

Patent
The condition of being open or unblocked (as with the fallopian tubes)

Pelvic Inflammatory Disease (PID)
An infection of the pelvic organs that may lead to tubal blockage and pelvic adhesions

Pituitary gland
The pituitary gland is situated in the brain near the hypothalamus and controls all hormonal functions, including the gonads, adrenal glands and thyroid gland.

Polar body
The discarded genetic material from the final female germ cell division. When an egg matures it discards half of its genetic material so that the egg cell can fuse with the male genes inside a sperm. This discarded genetic material can be seen in the egg as a small round piece of cytoplasm at the edge of the egg cell. The presence of a polar body tells us that the egg is mature.

Polycystic ovaries
A condition found in women who don't ovulate; characterized by excessive production of male sex hormones (androgens) and the presence of cysts on the ovary. PCO can be without symptoms although some women who do show symptoms are said to have PCOS.

Polycystic Ovary Syndrome (PCOS)
A condition where the symptoms of having polycystic ovaries are evident. PCOS symptoms may include weight gain, acne and excessive hair growth.

Pregnancy (biochemical)
A positive pregnancy test that shows no evidence of a gestational sac or viable fetus on ultrasound. Can be classified as a very early miscarriage.

Pregnancy (clinical)
A pregnancy documented by ultrasound that shows a gestational sac containing a viable fetus in the uterus.

Progesterone
The hormone produced by the corpus luteum after ovulation that supports the development and maintenance of the uterine lining.

Pronuclear Stage (2PN)
A fertilized egg (zygote). In IVF, the pro nuclear stage is the morning after the egg collection.

Pronuclei (PN)
Two spherical structures seen in the middle of the egg 16–22 hours post insemination indicating fertilization. One of the spheres contains the female genetic material and the other contains the male genetic material before they fuse to form the genes of the embryo

R

RESOLVE
A national, nonprofit consumer organization offering education, advocacy, and support to those experiencing infertility

Retrograde ejaculation
A male infertility problem in which sperm travel into the bladder instead of out of the penis. Medical intervention is necessary in order to conceive

Robertsonian translocation
A genetic disorder where the long arms of two chromosomes fuse together and the 2 short arms may be lost. All the normal genetic material is present for the carrier but this condition causes multiple problems for the offspring. Occurs in about 1:1000 births.

S

Salpingectomy
Surgical removal of the fallopian tubes

Semen analysis
A laboratory test to assess semen quality, including sperm count, morphology, motility, semen viscosity and volume

Society for Assisted Reproductive Technology (SART)
An affiliate of the American Society for Reproductive Medicine composed of clinics and programs that provide ART. SART reports annual fertility clinic data to the CDC.

Sperm agglutination
Sperm clumping caused by antibodies or infection.

Sperm count
The number of sperm in the ejaculate

Sperm morphology
The quantity or percentage of sperm that appear microscopically normal.

Sperm motility
The ability of sperm to move and swim normally

Stillbirth
The birth of an infant with no signs of life after 20 or more weeks of gestation

Stimulated cycle
An ART cycle in which a women receives oral or injected fertility drugs to stimulate her ovaries to produce more follicles

T

TESA
Testicular Sperm Aspiration: Surgical removal of the sperm

Testes
The male reproductive gland; source of sperm and male sex hormones normally occurring paired in an external scrotal sac

Testicular biopsy
A minor surgical procedure used to take a small sample of testicular tissue for microscopic examination. Sperm may be retrieved using a testicular biopsy when there is a tubal blockage, which is preventing sperm from being ejaculated.

Testosterone
The male hormone

Translocation
When chromosomes are attached to each other or parts of the chromosomes have been switched with one another.

Tubal factor infertility
Structural or functional damage to one or both fallopian tubes that reduces fertility.

U

Ultrasound
A technique used in ART for visualizing the follicles in the ovaries, the gestational sac, or the fetus.

Umbilical Cord
Two arteries and one vein encased in a gelatinous tube leading from the baby to the placenta. The umbilical cord is used to exchange nutrients and waste between the mother and the developing baby.

Unbalanced translocation
When there is extra or missing material in the chromosomal makeup.

Unstimulated cycle
A type of ART cycle in which the woman does not receive drugs to stimulate her ovaries to produce more follicles. Instead, follicles develop naturally. She may produce only one or two follicles.

Urologist
A physician specializing in the genitourinary tract

Uterine factor
A disorder in the uterus (e.g. fibroid tumors) that reduces fertility

Uterus
The hollow muscular organ in which a fetus develops during pregnancy

V

Vagina
The canal leading from the cervix to the outside of a woman's body

Varicocele
A dilation of veins that carry blood out of the scrotum, which leads to elevated scrotal temperature; a major cause of male infertility

Vas Deferens
The tubes sperm move through from the testicles (epidermis) towards the seminal vesicles and prostate gland. These tubes are closed during a vasectomy.

Vasectomy
A surgical procedure to block sperm from being released in the ejaculate. Vasectomy is used as a form of birth control.

X

X Chromosome
The female chromosome; females have two X chromosomes in their genotype.

Y

Y Chromosome
The Male chromosome; males have one Y and one X chromosome in their genotype.

Y Chromosome Micro-Deletion
A genetic condition when a man has small parts of the Y chromosome missing. This can manifest itself in a low sperm count and is detectable with a blood test.

Z

Zona Pellucida
The outer 'shell' surrounding the egg which serves two purposes; firstly allowing only one sperm to enter the egg for fertilization and secondly to hold the cells of the developing embryo together before compaction.

Zygote
A fertilized egg that has not yet divided

About the Author

After receiving her PhD in Cell Biology studying Embryo Implantation at the University of Manchester, Rebecca continued her training at St. James' Hospital in Leeds, England. She spent ten years as a Senior Embryologist at Oregon Reproductive Medicine and is currently the Scientific Director of Fertility Solutions Sunshine Coast in Queensland, Australia.

Her many years working with patients inspired her to create a guide to help couples understand their options, and the tests, procedures, background, potential outcomes and the miracles that occurs in the hands of gifted and dedicated fertility health care professionals the world over.

The results of her efforts have contributed to the healthy births of more than 2000 babies to warm and loving families from all over the world.

Rebecca lives with her husband and two sons on the Sunshine Coast of Queensland, Australia.

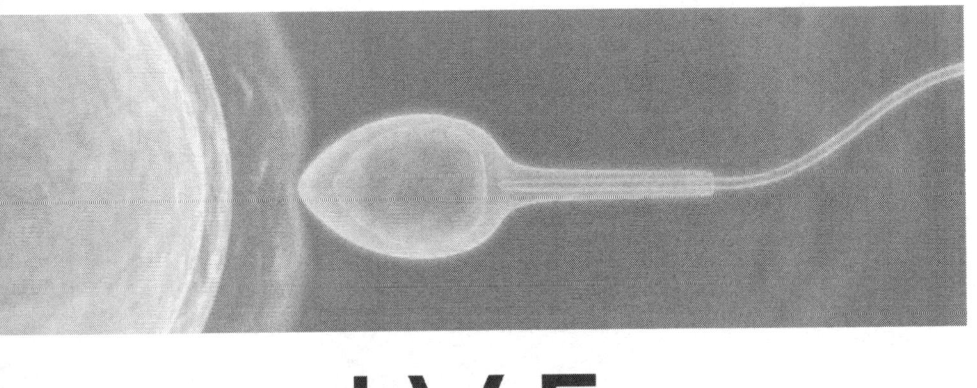

IVF

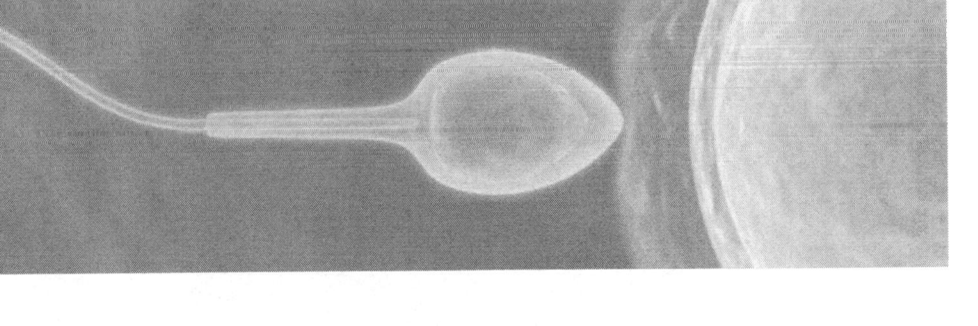

A PATIENT'S GUIDE

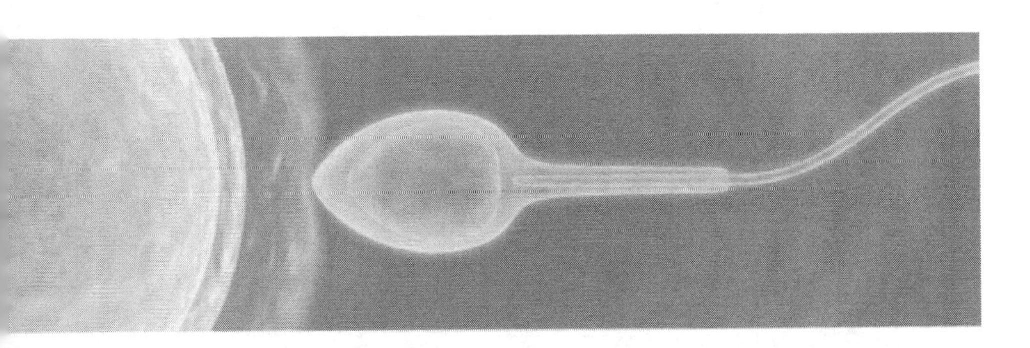

Rebecca Matthews PhD

21754705R00106

Made in the USA
Middletown, DE
11 July 2015